THE COMING POPULATION CRASH

ALSO BY FRED PEARCE

Confessions of an Eco-Sinner
Earth Then and Now
With Speed and Violence
When the Rivers Run Dry
Deep Jungle
Keepers of the Spring
Global Warming
The Dammed
The Big Green Book
Green Warriors
Climate and Man
Turning Up the Heat
Acid Rain
Watershed

The Coming Population Crash

And Our Planet's Surprising Future

FRED PEARCE

Beacon Press, Boston

Beacon Press
25 Beacon Street
Boston, Massachusetts 02108-2892
www.beacon.org

Beacon Press books
are published under the auspices of
the Unitarian Universalist Association of Congregations.

13 12 11 10   8 7 6 5 4 3 2

This book is printed on acid-free paper that meets the uncoated paper ANSI/
NISO specifications for permanence as revised in 1992.

Design and composition by Wilsted & Taylor Publishing Services

Library of Congress Cataloging-in-Publication Data

Pearce, Fred.
  The coming population crash : and our planet's surprising future /
Fred Pearce.
    p. cm.
  Includes bibliographical references and index.
  ISBN 978-0-8070-8583-7 (hbk. : alk. paper) 1. Population—Social aspects.
2. Population forecasting. I. Title.
  HB849.44.P43 2010
  304.601'12—dc22

2009038652

to Jack Caldwell,
most humane of demographers

The Coming Population Crash

**How boom is turning to bust. Half the world's women
are having too few babies to sustain present populations.**

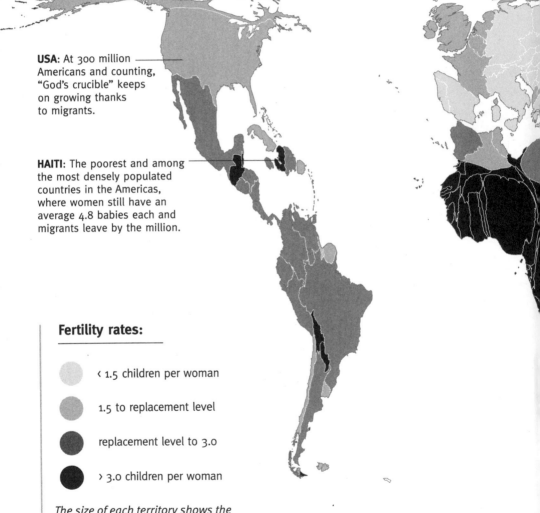

USA: At 300 million
Americans and counting,
"God's crucible" keeps
on growing thanks
to migrants.

HAITI: The poorest and among
the most densely populated
countries in the Americas,
where women still have an
average 4.8 babies each and
migrants leave by the million.

Fertility rates:

‹ 1.5 children per woman

1.5 to replacement level

replacement level to 3.0

› 3.0 children per woman

*The size of each territory shows the
relative proportion of the world's
population living there.*

*NB: Replacement level varies with each country, from 2.04 (Réunion)
to 3.35 in AIDS-ravaged Swaziland. World average is 2.3.*

EUROPE: With fertility below 1.5 from Madrid to Moscow, fears are growing of a demographic implosion. The last couple of years have seen a slight rise, but will it persist?

IRAN: Mullah power need not mean high fertility. In contrast to its neighbors, Iran has a fertility rate of a below-replacement 1.7 children per woman.

CHINA: Official figures say fertility in the world's most populous country is 1.8 children per woman, but unofficially it looks like 1.2.

INDIA: Fertility is falling fast, but before India's population stabilizes there will be more Indians than Chinese.

PHILIPPINES: The world's largest supplier of nurses and sailors is one place where the Catholic Church still imposes its grip. With abortion banned and contraception ostracized, fertility remains at 3.3.

SOUTH AFRICA: Women have an average 2.4 children today. Given the death rate from AIDS, that is below replacement level. The country could soon be contracting.

AUSTRALIA: Aussie women are still on childbirth strike, despite cash baby bonuses and ministerial exhortations to have "one for Mum, one for Dad, and one for the country."

Contents

I was brought up in Lenham, a village in the middle of Kent in England. On my final day at the village primary school, I walked down Honeywood Road, where I lived, and past the imposing Honeywood House to the vicarage, where I received a small cash award, the Honeywood Prize. It paid out from a trust set up more than three hundred years before by the village's most famous inhabitant, Mary Honeywood. The money was meant to buy tools for whatever trade I chose: that of a carpenter, say, or a blacksmith. So I bought a few pens and a compass and protractor for my future studies, and forgot all about it. Only later did I discover that Mary Honeywood is most famous—not for her house or her road or her prize, but for her fecundity. When she died in 1620 at the age of ninety-three, and her body was placed in Lenham churchyard, Mary Honeywood left behind 16 children, 114 grandchildren, 228 great-grandchildren, and 9 great-great-grandchildren—a total progeny of 367. So far as I can establish, that remains a world record for one woman in her lifetime. I suspect it is a record that will stand forever.

Demography is destiny. But not always in the ways we imagine. It underlies much of our world, shifting the tectonic plates on which our civilization is built. Never has that been more true than today. Wherever we look, population issues are among the most toxic headline-grabbers. From Gaza to Grozny and Rwanda to Afghanistan, baby booms are blamed for war and genocide. Festering slums burst into tribal violence in Kenya. Teenage terrorists lurk in refugee camps and overcrowded madrassas. Migrants from poor, overbreeding states are flooding Europe and North America. Overpopulation is the unspoken driver of environmental destruction. Millions of environmental refugees will soon be fleeing from spreading deserts and drowned deltas, as China's billion-plus inhabitants undermine all efforts to halt climate change.

The stats seem scary, too. The world's population is approaching seven billion—four times what it was a century ago. Never have there been so many mothers, and with half the people in some countries under sixteen years old, there are billions more baby-makers in the pipeline. Meanwhile, the world's masses are on the move. Some 200 million people go to bed in a country different from the one they were born in.

No wonder the language is bleak. Dickensian. Malthusian. It's Sodom and Gomorrah. It's Armageddon. We fear an overpopulated world teeming with the dispossessed and the alienated, the fanatical and the fascist, the wetbacks and the snakeheads, the Humvee-driving superpolluters and the dirt-poor deforesters. Surely we are racing to demographic disaster.

And yet slamming on the brakes seems almost as dangerous. For meanwhile, the insurgency of the old is looming. We are all living longer, healthier lives. Life expectancy has doubled since the 1950s. Back when I was born, 150 babies out of every thousand died before their first birthday. I could have been one of them. Now only fifty die. Should we cherish or fear this? Is good luck for the world's babies bad luck for the planet? It is sometimes said that more than half of all the people who have lived on the earth are alive today. This is nonsense. Just under 7 billion of the total human roll call of 100 billion are alive today. But what may well be true is that half of all the people who have ever managed to reach the age of sixty-five are alive today.

But don't despair. There is something you may not have guessed—something that may save us all. The population "bomb" is being defused. Only gradually, because the children of the greatest population explosion in history are still mostly of childbearing age, but it is happening. They may be having seven children in Mali, and six in Afghanistan, but half of the world's women are now having two children or fewer—not just in rich countries, but in Iran and parts of India, Burma and Brazil, Vietnam and South Africa. Mothers today have fewer than half as many offspring as their own mothers. This is happening mostly out of choice and not compulsion. Women have always wanted freedom, not domestic drudgery and the childbirth treadmill. And now that most of their babies survive to adulthood, they are grabbing it.

This book is the story of the peoplequake, the dramatic convulsion of the world's population that began with the Industrial Revolution and continues today. It is the story of how the tectonic plates of human population are shifting, and what this means for us and future generations. We see those plates shifting in the mosques of Iran and the slums of Mumbai; the vodka shops of Moscow and the killing fields of Rwanda; the demographic battlegrounds of Israel and the laid-back saunas of Stockholm.

If you are over forty-five, you have lived through a period when the world population has doubled. No past generation has experienced such an era—and probably no future generation will either. But if you are under forty-five, you will almost certainly live to see a world population that is declining—for the first time since the Black Death, almost seven hundred years ago. And it may happen soon. Demographers predict that, as during the Great Depression of the 1930s, the global recession that began in 2009 will encourage people to give up on babies for the duration.

The future will be a different world in other ways, too. The average citizen of the world today is under thirty. Before he or she dies, the average will probably be over fifty. In parts of aging Europe, there are already fewer than two taxpaying workers to support every pensioner. Book your places now for the global old folks' home. But the quake is not just about numbers. It is about age and sex and women's rights and war and migration; about the rise and fall of nations and, some fear, the end of the family. It is about environmental limits and climate change and the fertility of soils and minds.

The story opens with Bob Malthus, a morose eighteenth-century vicar spooked by two revolutions—the French and the Industrial—counting his stunted parish flock and imagining our demographic doom. From the Irish potato famine to Rwanda, the story follows the evolution of Malthusian fears of overpopulation. It tracks the terrifying logic of the twentieth-century eugenics scientists and the concerns of their birth-controlling successors who imposed coercive family planning in China, India, and elsewhere. It catches up with the new century's migrants and refugees and pensioners and, diminishing in number though they are, the babies of our planet.

It explores how demography drove the rise of the Asian tiger economies and China's economic miracle—and how it will soon undermine them both. It charts shrinking Europe, and how by midcentury Russia could have fewer people than Yemen. It follows the declining power of Catholic and Islamic clerics alike to lay down the law in the bedroom. It takes the political temperature of the "youth bulge" creating mayhem in the Middle East.

Most of all, it investigates the baby boom generation, born during the late twentieth century as world birth rates for a while reached double the death rates. The baby boomers are now adults, driving the global economy. But soon they will grow old.

And as the baby boomers start to die, global deaths will exceed global births. One way or another, their fate will be the fate of us all, for the boomers changed the planet. They were born into a world of resource abundance and will leave behind a world of profound resource scarcity. They brought us peak population, and with it peak oil and peak mining and peak trade and peak pollution. They will leave behind peak temperatures, too.

Have they done so much damage to the planet that the worst environmental nightmares are about to come true? Was the British government's chief scientist right to say in early 2009 that we face a "perfect storm" of food, energy, and water shortages by 2030? Was the Gaia scientist Jim Lovelock right to argue in his ninetieth year that the result will be "death on a grand scale from famine and lack of water . . . a reduction to a billion people or less" by 2100?

Many believe so. But haven't we heard such fears before? Malthus, of course. But also William Vogt, the forgotten hero of the environmental movement, who captured the world's attention with similar warnings in 1948, and Paul Ehrlich, whose *Population Bomb* repeated those warnings in 1968. None have come true—yet. So could the techno-optimists be right that our ingenuity will see us through to a new age? This is the first time in history that we have been able to foresee with some certainty a decline in our numbers. It means that if we can accommodate the imminent population peak, survival on planet Earth ought to become easier. That is not a cause for complacency. There are some choppy waters ahead, for sure, especially over climate change. We will need all our ingenuity to get through that, and to find ways to feed the eight or nine billion people who will inhabit the earth by 2040. But it could be a cause for hope as well. Optimism, even. Should we look forward to the benefits of a return to center stage of the tribal elders? Might the final legacy of aging boomers be a greener, happier, and more frugal world?

PART ONE

·

MALTHUSIAN NIGHTMARES

The shy young cleric with the cleft lip was an unlikely revolutionary. But two hundred years ago, "Bob" Malthus was the first Westerner to warn about rising population, which he said would end in pandemic, war, and famine. His doom-laden vision became a foundation stone of British imperial policy round the world, including its response to the Irish famine of the 1840s, in which more than a million died. And it underpinned the rise of one of the twentieth century's most toxic intellectual movements: eugenics. Today, many environmentalists see Malthus as their founding father. But in doing so, they have company that may surprise them.

THE COMING POPULATION CRASH

CHAPTER I

A Dark and Terrible Genius

The lane up the hill to Okewood Chapel is lined with bluebells and wild daffodils. An overpowering scent of garlic comes from the hedgerow. The little thirteenth-century church is set in a small clearing in the middle of woodland on the gentle Surrey downs south of London. It is a place of worship without a village, as it was in 1789 when the young Reverend Thomas Robert Malthus became its vicar.

Malthus lived with his parents a few miles away and rode across the hills to preside over baptisms and funerals. The workload perplexed him. In three years, he provided the religious rites for fifty-seven births and only twelve deaths. And there was something else. "It cannot fail to be remarked," he noted, "that the sons of labourers are very apt to be stunted in their growth. Boys that you would guess to be 14 or 15 are upon enquiry, frequently found to be 18 or 19." The local youths were, as Malthus's biographer Patricia James put it, "a different race from the lads who played cricket at Cambridge," where Malthus had a part-time university fellowship. The Surrey rustics lived almost entirely on bread and potatoes, not venison and fish.

The young Reverend Malthus was a recluse—unmarried, shy, with a pronounced lisp from a cleft lip. But his father, Daniel, kept him engaged in the affairs of the day. The French had just had their revolution, overthrowing Louis XVI, and the world-famous Norfolk revolutionary Thomas Paine had published his liberation manifesto, *The Rights of Man*. Freedom was in the air. Daniel corresponded with French revolutionaries and had once brought the libertine philosopher

Jean-Jacques Rousseau and his mistress to the family home. He was also a friend of the English journalist William Godwin, who in 1793 published a popular manifesto for an anarchist utopia called *Enquiry Concerning Political Justice*. In it, Godwin predicted an "eclipsing of the desire for sex by the development of intellectual pleasures." Childbirth would end. The world would become "a people of men and not children. Generation will not succeed generation. There will be no war, no crimes, no administration of justice, and no government. Every man will seek, with ineffable ardour, the good of all."

Malthus, who was known to family and friends as Bob, was having none of this libertarian babble. In evening discussions with his father, he laid into Godwin's optimism with particular contempt. And into the night, he was working on his own retaliatory text, an exercise in pessimism based on his experiences with his country parishioners. The result was his *Essay on the Principles of Population*, first published in 1798. Godwin's notion of a utopian world without sex and procreation was nonsense. In the real world, said Malthus, overpopulation was a looming threat, because the toiling masses were on a treadmill of sex and procreation. The poor reproduced in such numbers that eventually they would be culled by hunger and disease, he said. This was nature's way of restoring a balance in human population.

Bob Malthus's vision was simple, bleak, and devastating. Human population would, until checked, always grow exponentially. Just as a couple might produce four children, who would in turn create eight, and so on, a city of one million people would become two million in a generation, then four million, eight million, sixteen million, and so on. But food production could never grow so fast. At the most, it could grow arithmetically, able to feed one million, rising to two million in the next generation, then three million and four million and so on. The masses would inevitably run out of food or suffer disease till deaths brought the population back to its former level. It was, he said, happening in his own parish as births outnumbered deaths and the young grew up stunted. The whole country was going the same way. Probably the whole world.

Nothing could be done, he believed. All efforts to make the poor better off, or to relieve their suffering, would fail. Charity would only encourage more births, leading to a yet more calamitous population crash. To suggest otherwise was "an unpardonable deceit on the poor."

Malthus saw this gloomy prognosis as a natural state of affairs, but one with political consequences. It led him to oppose the English Poor Laws, which had for two hundred years offered the destitute meager protection inside workhouses. He said they simply encouraged early marriage and subsidized large families, and should be repealed.

For Malthus, this was a moral argument too. "Dependent poverty ought to be held disgraceful," he said. "A man who is born into a world already possessed, if he cannot get subsistence from his parents, or if the society do not want his labour, has no claim of right to the smallest portion of food, and, in fact, has no business to be where he is. At nature's mighty feast there is no vacant cover for him. She tells him to be gone."

This was cruel and haughty moralizing from a man of wealth. But he must have felt events were vindicating his pessimism. The public mood was changing. Idealism and utopias were out. The French Revolution had degenerated into a "reign of terror" under Robespierre, a disciple of Rousseau, his dad's old libertine friend. And his adversary Godwin's life had gone off track after his wife, the early feminist Mary Wollstonecraft, died. Their daughter had eloped at sixteen with the poet Shelley and wrote the far-from-utopian gothic horror novel *Frankenstein*.

Just as Godwin had earlier caught the wave of revolutionary optimism, so Malthus now rode the backlash. Malthusianism, as his bleak theory swiftly became known, was the talk of London's salons. The shy vicar soon counted many of the great mill owners and employers of the poor among his friends. He turned the essay into a full-length book, and over the next three decades he regularly brought out new editions, commenting on issues of the day, toughening his stance against the Poor Laws, and taking the bosses' side in industrial disputes in the fast-growing manufacturing towns.

Rising population was no distant threat, he told his growing audience. Malthus's natural law of overpopulation was being acted out there and then. It "exists at present over the greatest part of the globe" and "with few exceptions has been almost constantly acting upon all the countries of which we have any account." Not only were there limits to future growth, but the world was already operating at its limit. In a sentence that could come from a modern eco-tract, Malthus wrote: "The power of population is infinitely greater than the power in the earth to produce subsistence for man."

His influence soon extended far beyond British shores. In 1805 Malthus took up a teaching post at the college of the East India Company in Hertfordshire. The company was a capitalist colossus, conducting huge amounts of trade with the Far East and running much of India on behalf of the Crown. Malthus was made the world's first professor of political economy. For thirty years, until his death, he taught future administrators of the British Empire about the perils of overpopulation and the pointlessness of charity. As the modern Australian demographer Jack Caldwell puts it, "Malthus ensured that generations of British officials and scholars saw [the world] in Malthusian terms." His students included Charles Trevelyan, who, as we shall see, later presided with Malthusian hard-heartedness over the Irish potato famine in the 1840s.

. . .

Two hundred years ago, as Malthus baptized new souls in rural Okewood, Britain was becoming the workshop of the world. James Watt's steam engine created an industrial revolution that transformed the country into a manufacturing powerhouse unlike anything the world had ever seen. Optimists rejoiced at the brave new world, but Malthus saw his pastoral world disappearing. While many hailed the dawning of an age of prosperity, he saw a new nightmare amid what the poet William Blake called England's "dark satanic mills." The population of England doubled between 1750 and 1800 and then doubled again, reaching 24 million by 1830.

Most of the new population was manning the factories in the new industrial cities. The population of Manchester, the epicenter of the Industrial Revolution and the first manufacturing city in the world, increased sixfold from 1770 to 1830. Not far behind were Bradford and Leeds, the homes of wool manufacture; the steel city of Sheffield; Birmingham, the "city of a thousand trades"; and Liverpool, the gateway to the Empire. London had just become the largest city in the world, with a million inhabitants, and by midcentury its population would exceed two million.

In these industrial infernos, people choked in the foul air and drank water carrying typhus and cholera. As epidemics ripped through the overcrowded alleys and back-to-back houses, more than a quarter of the children born in this period were dead before they reached five years old, and another quarter before their teens were out. Life expectancy among the new laboring classes was below twenty years. Edwin Chadwick, in his famous 1842 *Report on the Sanitary Condition of the Labouring Population of Great Britain*, found that urban professionals and their families lived to thirty-eight years, tradesmen to twenty years, and "mechanics, labourers and their families" to an average of just seventeen years.

So why was the population growing? The answer seems to have been an enthusiasm for more babies. In the industrial towns, a typical woman had six children, compared to the historical rural average of four or five. With jobs in the factories and a run of good harvests, people wanted more children and reckoned they could feed them. Whatever Malthus thought, this was no blind rush to oblivion. This was rational calculation—not least because Malthus's new friends, the industrialists, were keen to take on this abundant workforce.

The laws of supply and demand did not work perfectly, however. By the 1830s, the industrialists were laying off workers, and poor harvests left a tenth of the country's population on Poor Law handouts. Malthus at last believed he saw his natural law at work. He led a growing campaign against a Poor Law that "subsidised marriage between pauper boys and girls." He backed the Poor Law Amendment Act, which became law in 1834, four months before his death. It greatly reduced relief and made workhouses "as like prisons as possible." It "forced the poor to emigrate, to work for lower wages, [or] live on a coarser sort of food," as one parliamentarian put it.

The masses called it Malthus's Law. It sealed his reputation. "Population Malthus" was the man who wanted to stop the poor from marrying. The humanist William Hazlitt accused him of "starving the children of the poor to feed the horses of the rich." His old adversary Godwin called him "a dark and terrible genius that is ever at hand to blast all the hopes of mankind."

He had a new enemy too. Charles Dickens saw in Malthus a ready target for his campaigning novels. In *Oliver Twist*, first published in 1837, Oliver's pathetic request for more gruel in one of the new supertough workhouses was a clarion call against Malthusianism. In *Hard Times*, the mean-spirited industrialist Thomas Gradgrind had a child called Malthus. In *A Christmas Carol*, the tight-fisted Ebenezer Scrooge was a satire on Malthus; Scrooge said of those who would rather die than go to the workhouse, "They had better do it, and decrease the surplus population" (a phrase, incidentally, that made it into a modern scabrous satire, the cartoon *South Park*, through the mouth of Cartman).

Malthus's ideas continued to cause controversy after his death. Europe's leading revolutionary, Karl Marx, called them "a libel on the human race," and the man himself "a sycophant of the landed aristocracy, bent on building the capitalist case for the inevitability of poverty." Charles Darwin said he had his flash of inspiration about the survival of the fittest in the natural world after reading Malthus. "I had at last got a theory by which to work," he said.

But while Darwin's world view is now universally held, Malthus still informs and enrages on issues from climate change to world poverty. The Surrey vicar's natural law underpins green fears about "limits to growth." Yet one recent reviewer called Malthus "an intellectual ancestor of the neo-conservatives of the present."

Was Malthus right? He was certainly correct to pinpoint population as a potent economic force for the first time. Developing a theory about the interplay of resources and population made him a towering figure. But he was clearly wrong that nineteenth- and twentieth-century Britain could not safely have a growing population. There are today more than twice as many Britons as when Malthus died. They are richer and healthier and live longer lives. And as he was wrong for Britain, so he was wrong for the world.

The immediate failure of his prediction was partly bad luck. Malthus was writing at the end of a thousand years during which European death rates had been largely determined by harvests. But far from catapulting the continent to disaster, as Malthus feared, the Industrial Revolution changed the rules of the game. It dramatically increased what environmentalists today call the "carrying capacity" of the planet. Britain ceased to grow its own food and instead fed its soaring population from newly colonized lands. It brought sugar from the Caribbean, wheat from India, tea from Ceylon, and meat from Australia—all secured with homemade armaments and paid for with manufactured goods.

Malthus didn't see that technology could make a nonsense of his natural law. But just as important, I think, he was wrong about human nature. He saw the poor as mindless beasts driven by crude natural forces, incapable of controlling their fertility. That was his "libel" on humanity. And it rather ignored the fact that his subjects were already controlling their fertility.

Even today, we often assume that before modern contraceptives, every sexual act was a potential pregnancy. It is true that short of entering a nunnery, no methods had the kind of success rates achieved today. Nonetheless, in eighteenth-century Britain, it was perfectly possible for women to keep their live births to, say, three or four rather than six. But Malthus was brought up at a time when contraception was rarely discussed. A retiring bachelor, sitting in his father's Surrey drawing room, he was simply ignorant of such matters.

· · ·

Greek, Roman, Ancient Egyptian, and Arabic literature all discuss means of contraception, from crocodile-dung pessaries to barriers and postcoital sponges and potions like honey and pepper. Many herbs used then are now known to contain chemicals that act as contraceptives, induce menstruation, or are outright abortifacients. The sap of a giant fennel known as silphium, which grew wild in what is now Libya, was so widely used in ancient Greece and Rome that it appears to have been driven to extinction. But the fourteenth-century inquisition records of the Pyrenees village of Montaillou have a local priest telling his mistress that "I have a certain herb. If a man wears it when he mingles his body with that of a woman he cannot engender, nor she conceive." Medical historian John Riddle reckons that the herb was a close relative of silphium. A wildflower known as Queen Anne's lace is still used in some parts of India, where unripened papaya is another old standby.

Withdrawal was widely advised in ancient times. Jewish writers describe it as "threshing inside and winnowing outside." The Peruvian Moche culture favored anal intercourse. Carvings on the temples of Angkor Wat in Cambodia show abortions. Riddle says such methods were widely used by women until the late Middle Ages to manage childbearing. After the Black Death in the 1340s, which left farm labor in short supply, the Church clamped down on contraception, and many old methods were lost or reduced to the status of witches' potions and old wives' tales. The witch hunts of this era were often tied up with panics about childbirth.

Contraception went underground, and some ideas fell out of use.

But there were still many social controls on childbirth—late marriages, for instance. Parish records in Colyton, Devon, show the mean age of marriage for women was twenty-seven years in the sixteenth century, and a hundred years later had risen to thirty—this at a time when life expectancy was only thirty-seven years. Women also lengthened the gap between children by continuing to breast-feed their infants, which suppresses ovulation by releasing the hormone prolactin. And there were always fallbacks like abortion.

Nineteenth-century Britain was riddled with hypocrisy about sex and birth control. William Cobbett wrote in 1829 that "farmers' wives, daughters and maids cannot now allude to, or hear named without blushing, those affairs of the homestead which they, within my memory, used to talk about as freely as of milking or of spinning." Church, state, and the medical profession, as well as males' romantic attachment to the sacred role of their wives as mothers, conspired to make sex impossible to discuss, contraception an unutterable evil, and fertility rates unmoved.

When investigative reporters at the *British Medical Journal* in 1868 replied to some of the numerous newspaper adverts aimed at ladies who were "temporarily indisposed," they found that this was a euphemism for pregnant and that more than half of the advertisers offered abortions. The *BMJ* was appalled. But when deprived of abortions, Victorian mothers often killed their unwanted newborns with opiates or gin, by starvation or smothering, or sometimes simply by abandoning them. The disposal of babies who survived the deaths of their mothers in childbirth was so prevalent that numerous philanthropists set up "foundling hospitals" to take unwanted children, with no questions asked. In France, Napoleonic foundling hospitals had a system of turntable devices at the doors, so parents were never even seen by those inside taking the child. A quarter of all Parisian babies were being abandoned at the hospitals.

The term "hospital" in this context was a euphemism. In practice the hospitals were clearinghouses from which the infants were sent off to "baby minders," where many died of neglect. A foundling hospital opened in central London by Thomas Coram, a sea captain and philanthropist, admitted fifteen thousand infants in its early years, of which ten thousand died. After visiting foundling hospitals in Moscow, Malthus said that "if a person wished to check population, and were not solicitous about the means, he could not propose a more effective measure."

As the survival rates of children increased, there was a growing demand for more reliable and effective birth control. France was the first European country to break the taboo on public discussion of how to prevent babies. Underground erotic novels such as Mirabeau's *Le rideau levé*, published in 1786, provided valuable instruction on withdrawal. Birth rates in France began to fall about then.

The United States was next, with France as the model. Early campaigner Robert Dale Owen wrote that "in France, men consider [withdrawal] a point of honour and learn to make the necessary effort." But American women did not want to rely on their men. They began buying syringes to insert a spermicidal douche, various formulas for which could be found in a booklet published in the 1830s by Owen and a New England physician called Charles Knowlton. To ensure sales, a French name for the potions was *de rigueur*. No doubt this is why the modern condom—one of the early practical results of the discovery, made in the 1830s by bankrupt and imprisoned American inventor Charles Goodyear, of how to vulcanize rubber—is often known as a "French letter."

The United States suffered a backlash against contraception after Anthony Comstock, a dry goods salesman who headed the New York Society for the Suppression of Vice, persuaded Congress in 1873 to ban its distribution. Meanwhile, as late as 1869, *The Lancet* was advising British doctors that the use of contraception in marriage meant the wife "acquired the condition of mind of a prostitute . . . and the husband was in effect practicing masturbation."

While the French relied on erotic novels and the Americans on doctors, the British were eventually indebted for their contraceptive enlightenment to a working-class tradesman and union activist. Francis Place, a tailor of leather breeches from Charing Cross in London who campaigned for universal suffrage, also took a keen interest in population issues. With commendable independence, he attacked Godwin for his utopianism, but also laid into Malthus for demonizing the poor.

In 1823 Place wrote leaflets and handbills extolling both withdrawal and the use of "a piece of sponge about an inch square, being placed in the vagina previous to coition, and afterwards withdrawn by means of a double thread or bobbin." To reassure the suspicious, he said the method was "getting fast into use among the working people of London."

Place was a serious campaigner. He toured the streets of London and northern industrial towns, flyposting his bills in working-class areas. There was outrage in the capital. One of his young (and very possibly libidinous) supporters was a seventeen-year-old boy called John Stuart Mill, who later became a leading political philosopher and member of Parliament. Mill was locked up for distributing the leaflet. But the message caught on better in the north, especially in textile mills, where women were by now widely employed. They became Britain's modern contraceptive pioneers.

Place was a prophet of contraception. But he was a prophet without honour in his own bedroom. He and his wife, who married in their teens, produced fifteen children, ten of whom reached adulthood. Largely as a result, they lived in constant poverty. Malthus, who professed that contraception was a sin, nonetheless contrived to produce only three children with his wife Harriet, whom he married in 1804. One died young and the other two, though married, never had children.

Malthus's genes may not have lived on, but his genius certainly did. So I was curious to find no mention in Okewood Chapel of its distinguished old vicar. Rather than a plaque commemorating one of the world's most influential thinkers, the chapel was festooned with "Make Poverty History" banners and posters extolling its parishioners' efforts to help the children of Joy Community School in Zambia. The Reverend Malthus would have been appalled at such misguided philanthropy. Was the current vicar, I wondered, engaged in a calculated snub? Sadly, not. The Reverend Nigel Johnson Knights told me he had never heard of the association between his church and Malthus.

CHAPTER 2

The Road to Skibbereen

West Cork is today a world of endless sandy bays washed by the Atlantic swell, where celebrities like movie mogul Lord Putnam, actor Jeremy Irons, and Ireland's richest man, the publisher Tony O'Reilly, all have homes. But there are ghosts here. Ghosts of the terrible Irish famine a century and a half ago, which killed a million people. And also of Bob Malthus. His natural law—that the poor will grow their numbers until cut down by disease or hunger—was said at the time to have caused the famine. But was it a case study in the operation of Malthus's law—or an illustration of its political misuse? In reality, the famine may be a terrible example of how, in the hands of mean-spirited politicians, Malthusianism can become a self-fulfilling prophesy.

The Irish famine lasted from late 1845 to 1849. It was Europe's worst modern peacetime human catastrophe. Reading about it today has the same numbing quality as reading about Darfur or Rwanda. "I entered some hovels," said the *Times* (of London) correspondent after a visit to the West Cork town of Skibbereen with the local doctor on Christmas Eve 1946. "In the first, six famished and ghastly skeletons, to all appearances dead, were huddled in a corner. I approached with horror and found they were still alive. In a few minutes, I was surrounded by at least 200 such phantoms. My heart sickens at the recital. In another house, within 500 yards of the cavalry station, the doctor found seven wretches lying unable to move. One had been dead many hours, but the others were unable to move either themselves or the corpse."

One teacher, watching his community disappearing, wrote: "The poor famine-stricken people were found by the wayside, emaciated corpses, partly green from eating dockweeds and nettles, and partly blue from cholera and dysentery."

A few weeks later, despite the publicity, things were no better in Skibbereen. The *Illustrated London News* reported that "not a single house out of 500 could boast of being free from death and fever." Many bodies were thrown into open trenches in the graveyard. In the nearby parish of Kilmoe, the undertaker improvised a coffin with hinged bottoms, so a corpse could be dropped into the grave and the box taken back for the next occupant. Eventually some ten thousand were buried in the famine pits of the Skibbereen workhouse, out of a parish population of twenty thousand. The local landlords, Lord Carbery and Sir William Wrizon-Becher, who between them drew £25,000 in rents a year, remained in their great houses as the famine unfolded.

So it was across western Ireland, as the potatoes on which the poor relied for sustenance were consumed by a virulent fungus that turned whole fields of them to putrefying mush within hours. *Phytophthera infestana* reached Ireland from Belgium in September 1845. Within three months it had destroyed a third of the crop. In the following two years, three-quarters of the crop rotted. A third was lost in 1848. A million died of hunger and disease during this plague, and after the famine a further million were shipped out by landlords anxious to be done with their tenants, or departed of their own accord for the New World. In half a decade, Ireland's population fell by a third, and eventually by half. Few lands have ever witnessed such a catastrophe; fewer still while under the control of the world's richest empire.

What caused this catastrophe? It wasn't just the fungus. Though all this happened a decade after his death, the shadow of Malthus's theories of overpopulation loomed large. The island's population, Malthusians of the time said, had grown beyond sustainable levels and crashed, as it inevitably must. The claim has some force. The potato had transformed Europe's ability to feed itself, and nowhere more than in Ireland. In the century before the famine, the island's population had been growing almost as fast as that of Britain. On the eve of the famine, it exceeded eight million, double what it was in 1780, with the poor south and west having seen the biggest rises. Ireland was the most densely populated country in Europe, with a third of the total population of the United Kingdom, of which it was then part.

But while Britain's industrial cities were soaking up labor and exporting products round the world, Ireland was still almost entirely rural. Its inhabitants mostly lived on 700,000 farms, nearly half of which covered less than two acres. Only potatoes could feed them from such meager resources. But feed them they did. Girls married at sixteen; boys at seventeen or eighteen. They had large families. Asked why Irish youths married so young, the Catholic Bishop of Raphoe told an inquiry in 1835, "They cannot be worse off than they are . . . and they may help each other."

Malthus did not in his *Essay* specifically predict disaster in Ireland. But in a typical mixture of insight and withering contempt, he noted that "the extended use of potatoes has allowed a very rapid increase [in Ireland's population] during the last century. . . . The cheapness of this nourishing root . . . joined to the ignorance and barbarism of the people have encouraged marriage to such a degree that the population has pushed much beyond the industry and present resources of the country."

During the famine, many sought a Malthusian explanation. One was the journalist Herbert Spencer, a fan of Malthus who anticipated Darwin by coining the term "survival of the fittest." He argued that people who breed too much take "the high road to extinction. This truth we have recently seen exemplified in Ireland." William Greg, one of Darwin's old student chums but by then a retired Lancashire mill owner, put it more crudely: "The careless, squalid, un-aspiring Irishman multiplies like rabbits."

Ireland had, as Malthus saw, become ridiculously dependent on the potato. A typical working man ate a staggering eleven pounds of the tuber a day—and little else. The potato was the only crop that could provide sufficient nutrition for the population on their small personal patches of land, while leaving the tenants free to provide labor for the big, often English-owned, estates. The Irish were the poorest people in Western Europe. Most lived in one-room turf cabins without windows or furniture. But usually there were surplus potatoes, which they fed to the chickens and pigs that often shared their cabins. In bad times, as the Irish economist Cormac O'Grada put it, the pigs starved and not the people. Or that was how it worked until the great famine.

But was this a Malthusian famine? Any society so dependent on a single cheap crop would have been vulnerable to the potato blight—

whatever its population. The more so if, as in Ireland, the landlords continued to export the only other foodstuffs. Ireland supplied most of England's food imports, especially beef. The island was engaged in what was probably the largest livestock export trade in the world. Yet nobody thought to divert that food to keep the Irish alive during the famine. Ports such as Limerick remained exporters. And just as starving Ireland sold its livestock, so did individual farmers. Many sold their pigs to pay the rent, even as their children starved. Meanwhile, the rich and middle classes, the traders and doctors, government employees and priests, fed well enough.

Some Irish Americans have charged that, far from being a Malthusian disaster, what happened was intentional "genocide." This is unfair. The British set up workhouses; their soup kitchens delivered three million cups of thin gruel a day at one point; and they instigated public works programs to employ the starving. But all these initiatives were grudging, and a small fraction of what local officials insisted was essential. The measures were wrestled from a penny-pinching government against a charged political climate of disdain, bordering on hatred, for the Irish.

There was willful ignorance in England about the real state of affairs across the Irish Sea. And officials assumed for too long that if there was a genuine need for food, then market forces would meet it. The problem with this laissez-faire argument was that the penniless had no money to buy food. As the famine's best historian, Cecil Woodham-Smith, reported of Skibbereen at the height of the famine, "on market-day September 12 1846, there was not a single loaf of bread or pound of meal to be had in the town." Market solutions didn't deliver food; they sent hundreds of Skibbereen children to the town workhouse, where more than half died.

The man in charge of all this was Charles Trevelyan, an upright and unyielding civil servant at the head of the British Treasury, who became virtual dictator of the relief program for Ireland. Trevelyan was an old India hand, educated at the East India College during Malthus's time there. He had absorbed the great man's moral and economic ethos. He argued that the famine was both a "mechanism for reducing surplus population" and "a direct stroke of an all-wise and all-merciful Providence."

Trevelyan, brother-in-law of the famous historian Thomas Macaulay, regarded himself as a reformer. He was a member of the Clapham Sect in south London, whose earlier high-minded, evangelical, and philanthropic members included the great slavery abolitionist William Wilberforce. But in imperial matters he was a pious and interfering hardliner, with both God and Malthusian logic behind him. As deaths multiplied in the freezing winter of 1846–47, Trevelyan wrote: "The great evil with which we have to contend is not the physical evil of the famine, but the moral evil of the selfish, perverse and turbulent character of the people. . . . We are in the hands of providence, without a possibility of averting the catastrophe. We can only wait the result."

Such views were not unusual. The prime minister, Lord Russell, whose brother owned large Irish estates, professed "a Malthusian fear about the long-term effects of relief." Lord Clarendon, a major landowner as well as president of the Board of Trade in London, said that "doling out food merely to keep people alive would do nobody any permanent good." The editor of the *Economist*, referring to the famine, wrote that paying any man more than he was worth "would stimulate every man to marry and populate as fast as he could, like rabbits in a warren."

Ireland's problem was that it had been run for centuries as a semi-feudal colony. Its best pastures were taken over by around two thousand large estates, mostly owned by Protestant settlers who raised beef for export to Britain and to feed slaves in their Caribbean colonies. This left the native Irish to cultivate, as best they could, tiny plots on poorer soils. Ireland's calamity was that, facing the results of their own actions and policies, the British denied responsibility and employed Malthus's dictums to explain how it was all inevitable and a consequence of the actions of the victims themselves.

Malthusian logic had a way of demonizing those whom it punished for their poverty. The landlords of Ireland had for decades encouraged large families as a source of labor. But by the time of the famine, they were modernizing their farming methods and needed less labor. So amid the famine in 1846, starving tenants were evicted for nonpayment of rents, often under the gaze of government soldiers. Some half a million were ejected in this way, often herded onto "coffin ships" bound for America.

In London, these evictions were widely regarded as the solution to the "Irish problem." "I would sweep Connaught clean" said the country's viceroy, Lord Clarendon, in 1848. Ireland required "systematic ejectment of smallholders and squatting tenants," said cabinet member Lord Palmerston. He had his own large estate in Sligo and had shipped two thousand tenants to Canada at the height of the famine. Up to a quarter of them died on the journey. Those who survived were left hungry, destitute, bewildered, and often nearly naked on the docks of Montreal, or incarcerated in "fever sheds."

The Victorian English working class showed little solidarity or compassion for their Irish cousins. They were brought up on a diet of stories about the feckless Irish. The Irish, they believed, were sexually promiscuous. Not so. The Irish married young, but were more monogamous than the English. The Irish were, according to the Cambridge historian Charles Kingsley, "white chimpanzees," and according to *Punch*, the "missing link between the gorilla and the negro." This simian allusion may have its origins in widespread reports that starving Irish children grew hair on their faces, making them "look like monkeys." More likely these were the faces of starving, emaciated adults.

The Irish were regarded as traitors and spongers, too. The *Times*, whose own reporters had conveyed the sickening scenes in Skibbereen, nonetheless thundered in 1847 that "in no other country have men talked treason until they are hoarse, and then gone about begging for sympathy from their oppressors. In no other country have the people been so liberally and unthriftily helped by the nation they denounced and defied."

By early 1847, the Irish were no longer waiting to be thrown off their land. They were leaving in droves. Almost anyone from the countryside with the cash for the passage shipped out. Millions took steerage to North America during and after the famine. Another million or so sailed east to Liverpool, to work in the mills and on the railway construction projects of mainland Britain. The Irish diaspora had begun.

In the years after the famine, the Irish population continued to decline through migration and falling fertility at home. A Malthusian analysis might have forecast population recovery as pressure on the land diminished. But the exodus continued, and poverty remained as great as ever. As if to mock the Malthusian argument, famines returned even as the country emptied. It is difficult to escape the conclusion that the real cause of Ireland's destitution was not Malthusian natural laws but the unedifying laws of man.

Things have moved on since most of Ireland won its independence from Britain in the 1920s. The population has recovered—and wealth has grown. Castle Freke, the ancestral home of Skibbereen's Lord Carbery, has been an eerie ruin for almost a century. New money is abundant in the west of Ireland. The land's carrying capacity and the potential wealth of its people have proved far greater than Malthus or Trevelyan would have imagined.

In truth, the tragedies at Skibbereen and hundreds of other communities tell us more about the operation of the British Empire than about overpopulation. Given the government policies of the day, the fungus would have caused the Irish to starve even if there had been only a tenth as many of them. The potatoes on which they lived would have died anyhow. The alternative food supplies grown on Irish soil would have been exported just the same. But if events tell us little about Malthus's laws, they tell us a very great deal about the pernicious nature of his self-fulfilling nostrums about the folly of philanthropy. It is an evil that today's Malthusians might bear in mind.

CHAPTER 3

Saving the White Man

There is an intellectual thread that runs from Malthus and Darwin to modern thinking on the environment and the "population bomb." It passes through some dark places, like dreams of racial supremacy, forced sterilization of the "unfit," and the gas chambers of Nazi Germany. This is the world of eugenics, the study of methods of "improving" the human race that turned science into a Frankenstein monster. Eugenics is largely disowned now, but it was hugely influential during the first half of the twentieth century—in liberal and socialist circles as well as the political right. Some may see this chapter as an unpleasant detour. But eugenic thinking was important to many of the founders of modern population control and environmentalism, and lives on in some corners today.

Malthusian ideas gave birth to eugenics. The line runs through Charles Darwin, who said that his own theory of natural selection grew from reading Malthus's predictions about the effect of a "grand crush" in human populations. In a natural world of different species vying for resources, the crush would create competition in which only the fittest species survived. From there, it was a small step to the idea that similar conflicts could break out between human races or groups with particular genetic traits, and that if such competition was necessary for natural evolution, perhaps it was necessary for human well-being too.

The man who made that intellectual step was Darwin's half-cousin, Francis Galton. The heir of a Quaker arms manufacturer, he was a troubled child prodigy. After a mental breakdown in his early twenties, he abandoned a career in medicine to pursue ideas that took his fancy.

These included everything from meteorology (he produced some of the first weather maps) to African exploration (he helped chart what is today Namibia) and the science of human heredity. He saw everything from genius to drunkenness as inherited. Nature beat nurture every time. You might say, "He would see it that way, wouldn't he." It was very convenient and sociable for a man of such distinguished lineage and comrades to discern "many obvious cases of heredity among the Cambridge men who were at university about my own time."

Galton quantified some of these inherited traits, at any rate to his own satisfaction. His studies showed that distinguished men fathered distinguished sons. He did include daughters in his analysis, but concluded that inheritance on the male line exceeded that on the female line 70 to 30. He does not seem to have even considered that it might often be opportunity and privilege that were being passed on, through social rather than genetic forces.

Galton was a practical man. Having established his theory of inherited traits, he decided that, like racehorses, men could be bred to be champions. He wanted to improve the human race through selective breeding among the superior, while discouraging reproduction among the inferior. "What nature does blindly, slowly and ruthlessly, man may do providently, quickly and kindly." From the start, his agenda was racial. "Humanity shall be represented by the fittest races," he said. So he graded racial types according to their record in producing judges, statesmen, military commanders, scientists, artists, and so on. Anglo-Saxons and Nordics came out on top and "negroes" at the bottom, since, he said, "it is seldom that we hear of a white traveller meeting with a black chief whom he feels to be the better man."

Like Malthusianism, Galton's eugenics gained popularity among its originator's peers partly because it fed the vanity of the privileged and absolved them of responsibility for the rest. It drew, too, on what has been called the "dark side of Darwinism"—the idea that species might regress to more primitive forms. Galton thought this might happen to humans in a decadent welfare state, where the fittest no longer had a head start in the survival stakes.

Malthus thought that the middle classes had prudence and foresight and could determine their own futures, but that the poor were at the mercy of natural forces. Galton took this one stage further. The poor were a threat, not just through force of numbers but because their fertility would spread their genes far and wide, overwhelming the better genes of their superiors. As Darwin's old friend William Greg put it, the danger was that "in the eternal struggle for existence, it would be the inferior and less favoured race that prevailed . . . by virtue not of its good qualities but of its faults."

This became another reason for not helping the poor. Malthus sometimes opposed efforts to improve human health, such as Edward Jenner's smallpox vaccine, because they would boost population. But enthusiasts for eugenics went further, saying that vaccines stopped nature weeding out people vulnerable to its diseases. Dr. John Berry Haycraft told a meeting at the Royal College of Physicians on "Darwinism and Race Progress" in 1894 that "preventative medicine is trying a unique experiment, and the effect is already discernible—race decay."

This was nasty stuff. It was soon everywhere. Eugenics was the science of humanity in the machine age, and it eventually underpinned everything from better child care to compulsory sterilization. Humans were to be perfected by selective breeding. And nowhere, so far as Galton was concerned, did this matter more than in the world's top dog, Britain. "In no nation is a high human breed more necessary than in our own, for we plant our stock all over the world and lay the foundation of the dispositions and capacities of future millions of the human race," he told top Britons at the newly formed Sociological Society in London in 1904.

Some at the meeting quibbled. The science fiction author and socialist Herbert (H. G.) Wells pointed out that the sons of the elite may prosper through "a special knowledge of the channels of professional advance" rather than any genetic superiority. Genius could come from anywhere. William Shakespeare, after all, was from humble stock and produced no talented heirs. Wells also wondered how you distinguished the "fittest" from the rest. Galton wanted to prevent criminals from procreating. But Wells pointed out that "a large proportion of our present-day criminals are the brightest and boldest members of families living under impossible conditions." They might, he said, be brighter than "the average respectable person."

Wells was in a minority, however. The Irish playwright George Bernard Shaw caught the general mood of jingoistic eugenics when he said that "nothing but a eugenic religion can save our civilisation from the fate that has overtaken all previous civilisations."

In 1906 Galton wrote a novel about a eugenic utopia called *Kantsaywhere*. In hindsight, his athletic, militaristic young men and earthmother women resembled the Aryan master race of Nazi mythmaking. Sadly, the novel was not passed as of sufficient quality by any publisher and suffered the fate of the unfit, with its reproduction prevented.

But this was a rare setback. In 1907 the University of London opened a Francis Galton Laboratory for the Study of National Eugenics. And the intellectual tentacles of the new discipline spread. In an era in love with scientific socialism and the perfectibility of man, many on the left embraced eugenics in ways that would horrify later generations. Economist and Bloomsbury socialite John Maynard Keynes was a fan. The Fabian Society's Sidney Webb felt that eugenics could prevent Britain from "falling to the Irish and the Jews." Galton's protégé, the Quaker socialist Karl Pearson, wrote, "History shows me one way, and one way only, in which a high state of civilization has been produced, namely the struggle of race with race, and the survival of the physically and mentally fitter race." At the top of the heap, he saw "the Aryan success."

In 1912, a year after Galton's death, the first International Congress on Eugenics took place in London, presided over by Leonard Darwin, son of Charles and half-cousin of Galton. It was attended by former home secretary Winston Churchill; the prime minister, Arthur Balfour; the aging American inventor of the telephone, Alexander Graham Bell; and many others. Who, after all, could resist the invitation? Churchill's lineage was unparalleled, and he later, in a cabinet meeting, proposed sterilizing the "feeble-minded." The term "feeble-minded," incidentally, is generally credited to Charles Trevelyan, the Malthusian impresario of the Irish famine. It was a convenient catchall, used to cover petty criminals, prostitutes, vagrants, the mentally defective, and other undesirables. (It might once have been applied to Galton himself at the time of his youthful nervous breakdown, but let's draw a veil over that.)

The science of eugenics was sweeping the United States too, with terrifying practical consequences in the forced sterilization of those deemed unfit to reproduce. By 1926, seventeen states had active sterilization programs. One early victim, following a Supreme Court ruling, was Carrie Buck, a resident of the Virginia State Colony for Epileptics and the Feebleminded, whose only crime was having had a child after being raped.

Harry Laughlin, the director of the Eugenics Record Office, a new research center at the Cold Spring Harbor Laboratory, gave evidence in favor of the Carrie Buck ruling. And when legislators wanted to allow in more of the Nordic and Teutonic races of northwest Europe and cut out "inferior stock" from southern and eastern Europe, Laughlin agreed that the new laws would help prevent pollution of the national gene pool. Francis Walker, a former head of the U.S. census and president of the Massachusetts Institute of Technology, called the would-be migrants from the east "degraded . . . beaten men from beaten races . . . failures in the struggle for existence."

But there were also plenty of victims at home for this genetic witch hunt. Between the first and second world wars, some sixty thousand imbeciles, epileptics, and "feebleminded" persons were compulsorily sterilized in the United States, half of them in California. Other eugenic laws prevented the mentally ill from marrying and banned interracial marriage.

Eugenics was the face of population control in the first half of the twentieth century. It was a global movement, embraced by right and left alike. Imperial Japan forcibly sterilized tens of thousands of people diagnosed with mental illness, along with criminals judged to have a "genetic predisposition to commit crime" and victims of Hansen's disease (then called leprosy). Popular Japanese magazines carried articles on "eugenic marriage," inviting sweethearts to fill out questionnaires to determine whether their lovers were genetically fit. Meanwhile, pacifist Sweden sterilized sixty thousand of the mentally ill. The policy's leading advocates were Alva and Gunnar Myrdal, the intellectual founders of the welfare state.

Among its many shortcomings, eugenics was staggeringly sexist. Galton had regarded women as peripheral in passing on positive genetic traits. But they were uniquely in the firing line when it came to extinguishing damaging traits. More than 90 percent of those sterilized in Sweden were women. Their crimes were often as minor as joining a motorcycle gang or going to a dance hall, with the operation carried out on the say-so of a social worker or neighbor. It was as easy as getting an Antisocial Behavior Order today.

The British continued to support their intellectual baby. In 1931 three leading contributors to a series of BBC talks on the theme "What I would do with the world" supported sterilization of the unfit. They included a future minister for India, Leo Amery, who said such sterilization was needed to correct "short-sighted sentimentalism that encouraged the multiplication of the improvident and the incompetent." Britain never legalized enforced sterilization. Perhaps it never felt the need. Doctors often took the view of psychiatrist Carlos Blacker, secretary of the British Eugenics Society, that "defectives, being for the most part readily suggestible, should in most cases be easily persuaded."

In Germany, sterilization of the feeble-minded began in the late 1920s under Dr. Otmar von Verschuer. He became a mentor of Dr. Josef Mengele, the "angel of death" at the Auschwitz concentration camp who experimented on inmates. The Nazis sterilized about half a million people. In a Mother's Day radio address in 1935, Hitler's interior minister, Wilhelm Frick, railed at "how honest families had fewer children, while those less worthwhile and the bastardly multiplied uninhibitedly," and "how irresponsible German people coupled with alien races." Nazi policies were widely admired. U.S. eugenicists complained that "the Germans are beating us at our own game," and Britain's *Eugenics Review* noted approvingly how Germany's "gigantic eugenics experiment . . . is carried through without the slightest difficulty."

But the Nazis took their ideas of "racial hygiene" far beyond sterilization. Heinrich Himmler, head of the SS, first tried to fast-track propagation of the master race by paying "pure-bred" Aryan women to have unprotected sex with his officers. Then he began killing tens of thousands of the "unfit," including the handicapped, the chronically sick, and even the "demented elderly" (who were, you might have thought, hardly in a position to start breeding a new generation of demented elderly). All of which was a precursor to the Holocaust, the systematic extermination of millions of those whom the Nazis regarded simply as inferior races, such as Jews and Roma (sometimes called Gypsies), or otherwise beyond the pale, like Jehovah's Witnesses and homosexuals.

Enough has been written about the Holocaust, but what is worth remembering here is that it was an extreme expression of a more general eugenics movement whose boundaries extended far beyond Germany. Nor did its ideas die with the 1945 armistice.

The defeat of the Nazis in 1945 and the discovery of their "final solution" left eugenics disgraced in the public eye. The Universal Declaration of Human Rights, agreed upon at the newly created United Nations in 1948, banned compulsory sterilization. But much of the thinking, and some of the policies, lived on, and many former eugenics scientists resumed careers as anthropologists, geneticists, and social policy makers. When Verschuer, Germany's pioneer of racial hygiene, went to the United States in the 1950s, the American Eugenics Society welcomed him as an honorary member. Its head, Frederick Osborn—a man described by Wikipedia as "philanthropist, military leader and eugenicist"—was unrepentant at having written in 1937 that Verschuer's work rooting out hereditary diseases was "the most exciting experiment that had ever been tried."

It is fair to say that after 1945, the American Eugenics Society mostly promoted "positive" eugenics, encouraging babies from the favored, rather than the "negative" eugenics of sterilization. But the legacy of eugenics lingered. Osborn's other postwar activities involved founding the Pioneer Fund, whose promotion of "human race betterment" includes funding research into racial differences in IQ and running the Population Council, the foremost body for promoting family planning and population control worldwide.

Jack Caldwell, the leading postwar Australian demographer, says the concerns of the eugenics movement were a major force behind the development of population studies. Many leading demographers began as eugenics enthusiasts, and their work "established a precedent in terms of population scientists becoming involved with government attempts to control fertility." The idea of compulsory sterilization went on to obsess the postwar population control movement.

As we shall see, many of the twentieth century's most revered environmentalists also came from the same stable. Eugenics continues to color their fears that "overpopulated" megacities are breeding grounds for terrorism and that Europeans are losing out demographically to Africans and Muslims. Senior environmentalists still describe AIDS in Africa as a potential solution to the planet's Malthusian crisis, reducing population and weeding out the poor and feeble. Eugenics lives on.

· · ·

Most eugenics advocates were men. But not all. Margaret Sanger was the sixth child of a drunken socialist Irish stonecutter and his long-suffering wife, Anne, who went on to have five more children and died of tuberculosis at the age of forty-nine. Margaret Sanger always blamed her mother's death on the stress of, in all, eighteen pregnancies. Sanger's upbringing taught her two things, she said: that parents should have only the children they can support, and that a woman's life of bearing babies for a man who opposes contraception is not a happy one. "A free race cannot be born of slave mothers."

Sanger trained as a nurse, became politically active, and in defiance of the notorious Comstock Law of 1873, opened one of the United States' first birth control clinics, in Brooklyn in 1916. She launched *Birth Control Review* the following year and founded the American Birth Control League in 1921. It later became Planned Parenthood of America.

Sanger was a driven woman, a great talker, a tireless lecturer, and a brilliant hustler. She became the figurehead of the birth control movement in much of the world. One of her most important benefactors was Clarence Gamble, heir to the soap manufacturing empire Procter & Gamble. On his twenty-first birthday, Gamble received his first million dollars from the family fortune under the condition that he invest at least a tenth of it in some charitable enterprise. He chose birth control, and Sanger got the lion's share.

Sanger was a socialist and a feminist, but also a eugenicist and a Malthusian. There was a rod of steel in her desire to see the lower classes adopt family planning. She believed in reproductive rights but also social duty. Combining the thoughts of Malthus and Galton, she opposed organized charity as "the symptom of a malignant social disease." It was "the surest sign that our civilization has bred, is breeding and perpetuating constantly increasing numbers of defectives, delinquents and dependents." Rather than "shouldering the burden of the unthinking fecundity of others," society should prevent them from breeding. "More children from the fit, less from the unfit—that is the chief issue of birth control."

Sanger persuaded one of the United States' leading academic supporters of eugenics to become the first director of the American Birth Control League. Harvard history professor Lothrop Stoddard was a pacifist and a stamp collector, but also the author of books with such unedifying titles as *The Rising Tide of Color Against White World Supremacy* and *The Revolt Against Civilization: The Menace of the Under Man*. Stoddard saw evolution in the human as well as animal worlds as a process of "ever-increasing inequality." Social reform aimed at reducing inequality was "one of the most pernicious delusions that has ever afflicted mankind." *The Rising Tide* had a thinly veiled walk-on part in F. Scott Fitzgerald's *The Great Gatsby*. Football hero Tom Buchanan looks up from a book called *The Rise of the Colored Empires* and says, "If we don't look out, the white race will be utterly submerged. It's all scientific stuff; it's been proved."

Stoddard was not just a leading figure in U.S. eugenics and a co-founder of the birth control movement. He was also one of a trio of men from this era with a good claim to having founded twentieth-century American environmentalism. The others were Henry Fairfield Osborn, uncle of Frederick Osborn and president of the American Museum of Natural History, and a New York lawyer, hunter, and naturalist called Madison Grant. Together the three founded the Save the Redwoods League in 1917 to protect the redwood forests of Northern California. The organization is still in business today.

All three combined environmentalism with populist eugenics. Grant, who was for a time also vice president of the U.S. Immigration Restriction League, argued for Nazi-style selective sterilization to eliminate "the criminal, the diseased, and the insane, extending gradually to weaklings and perhaps ultimately to worthless race types." The year before raising his standard for the redwood, Grant had published his own rabble-rousing tome called *The Passing of the Great Race*, a study of the "racial history" of Europe and the threat posed to "the Nordic race." He believed hard northern winters had purged defective genes from the virile, blond-haired, blue-eyed, chivalric and warrior Nordic race. Their purity had to be protected. The book sold 1.6 million copies in several languages, including German, which translated "Nordic" as "Aryan."

Grant has a species of caribou named after him, and he helped create Glacier National Park. As secretary of the Bronx Zoo, he put on display in 1906 a Congolese pygmy called Ota Benga. The unfortunate man, who had been bought from soldiers of King Leopold of Belgium, occupied a cage next to a collection of apes to illustrate the stages of human evolution. When black clergy in New York tried to rescue him, the

New York Times said that to do so would be an affront to science. "The reverend colored brother should be told that evolution is now taught in the textbooks of all schools." After his eventual release, Ota Benga went into the Vermont woods and shot himself.

One of Grant's friends was former president and fellow hunter Teddy Roosevelt, who called *The Passing* "a capital book." John Burdon Sanderson Haldane, the top British geneticist, was also a fan. Several years before Grant's death in 1937, an aspiring young politician in Germany wrote to him. "The book is my bible," said Adolf Hitler. Two decades later, during the postwar Nuremberg trials, the head of the Nazi euthanasia program, Karl Brandt, also cited Grant's book as inspiration.

Today, environmentalists are generally seen as part of the political left. But conservation was for a long time the preserve of conservatives. It was a right-wing and often antidemocratic movement that regarded the poor masses as the instigators of environmental as well as racial destruction. Grant was far from alone in combining eugenics and environmentalism. Fellow top-drawer conservationists included Julian Huxley, a British biologist who later became the first director of UNESCO, the United Nations Educational, Scientific, and Cultural Organization, and founder in 1961 of the World Wildlife Fund. Huxley was a member of the British Eugenics Society for forty years, serving as president from 1959 to 1962.

The Huxleys looked like a one-family confirmation of Galton's ideas of inherited excellence. Julian's paternal grandfather, Thomas Huxley, was an early establishment cheerleader for Charles Darwin and widely called "Darwin's bulldog." And his elder brother was Aldous Huxley, author of *Brave New World*. The book describes a society in which humans are cloned to order in the laboratory, which Aldous himself described as a "foolproof system of eugenics."

While running UNESCO, Julian Huxley tried to keep the eugenic flame burning. He argued that "though any radical eugenic policy will be for many years politically and psychologically impossible, it will be important for UNESCO to see that ... the public mind is informed of the issues at stake, so that much that is now unthinkable may at least be thinkable." Otherwise, a future world might end up being inhabited by "the descendents of the least intelligent persons now living." Huxley proposed giving tax credits to encourage those most fit to reproduce—an idea taken up soon after in Singapore. And he was the first to suggest the idea of using sperm banks to shower the gene pool with DNA from Nobel Prize winners and others with high IQs.

But in his later years, Huxley's interest in eugenics declined and his concern for the planet's environment and its ability to feed itself grew. He was not alone.

PART TWO

.

RISE OF THE POPULATION CONTROLLERS

"It is probable that at least three-quarters of the human race will be wiped out . . . large areas of the Earth will have to be written off." The warning, made in 1948, has set the tone for fears about the population explosion ever since. It colored the cold war and kick-started a global campaign to prevent the poor of the world from breeding—a campaign that culminated in millions of forced vasectomies in India and China's draconian one-child policy.

CHAPTER 4

An Ornithologist Speaks

In 1917, at the height of the First World War, the small Belgian village of Passchendaele suffered what was at the time the largest bombardment of military hardware in history. Over five months, a million shells rained down on German, British, Australian, and Canadian soldiers dug into waterlogged trenches. They left three-quarters of a million dead. A few months before, a million had died in the Battle of The Somme. Overall, fifteen million died in the war. By the end, France had lost 3 percent of its population. Almost a tenth of British males between the ages of twenty and forty-five had perished. From the gleaming spires of Oxbridge universities to the backstreets of northern industrial towns, the sacrifice was huge. In the aftermath, a quarter of the country's young women wore poppies and did not marry.

Europe's fertility rates had been falling for a century. By the 1930s, this long-term trend, coupled with the impact of the wartime "lost generation" and a reluctance to have babies in an economic depression, left many European countries below the theoretical replacement level of just over two babies per woman. French women were having an average of 1.9 babies each, British women just 1.7. Paris responded by banning the promotion of family planning. The condom, known to many English speakers as a "French letter," was illegal in France. In Italy, Benito Mussolini declared a "demographic battle" to increase the population by 50 percent. He gave the country till 2050 to achieve this, but nonetheless demanded monthly reports from provinces to detect early backsliders.

In 1934 a feminist statistician in Britain, Enid Charles, published a book called *Twilight of Parenthood*. Later titled *The Menace of Under-Population*, it predicted widespread population decline in the West. The author blamed family planning. "The process of rationalising reproduction has produced a problem of the first magnitude," she warned. "The prosperous classes of industrial nations, like other ruling castes in the past, have become victims of their own ideology. . . . They have moulded the destiny of a civilisation that has lost the power to reproduce itself."

These fears of Europe's demographic demise stoked concern that the huddled masses of the rest of the world were on a demographic march that would, as the eugenicists warned, swamp the white races. The evidence for this was sketchy. In 1918 a worldwide flu pandemic killed an estimated forty million people and put back world population growth by an estimated five years. And Catholic theologians, with their worldwide network of priestly spies in confession boxes, were predicting a catastrophic population decline in poor countries as well as rich.

However, in 1928 Australia's former census chief, Sir George Handley Knibbs, concluded that the world's population had doubled in the previous century and was on course to grow to 3.9 billion by 2008 and 7.8 billion by 2089. Census returns from Europe's empires showed that innovations like sanitation, soap, and modern medicine were cutting death rates and spurring new population growth in poor countries. The French noticed the trend in North Africa and Indochina. The scrupulous Indian census revealed that as life expectancy rose and fertility stayed close to six children per woman, the population was growing by more than 1 percent a year.

Delight among imperial administrators at the growing health of their subjects was tempered by growing concern at their rising numbers. In 1939 a British Royal Commission said falling death rates meant that a commensurate reduction in the birth rate was "the most pressing need of the West Indian colonies." In 1944 the governor of Kenya described the "astonishing rate of increase" in its population as "the single cause to which the difficulties of the African people can be ascribed." And the French demographer Alfred Sauvy warned that rising numbers were driving independence movements in the French colonies.

Optimists dismissed these fears. The future Nobel Prize–winning American economist Theodore Schultz argued that "man has the ability and intelligence to lessen his dependence on cropland, on traditional agriculture and on depleting sources of energy, and can reduce the real costs of producing food for the growing world population. . . . Mankind's future is not foreordained by space, energy and cropland. It will be determined by the intelligent evolution of humanity." But Knibbs had called himself a "new Malthusian," and the prevailing view as the Second World War ended was one of concern that, outside Europe, a major global population surge would resume.

The immediate postwar world saw turmoil, with rationing and refugees in Europe, nuclear bomb tests in the Pacific and Arctic, and the start of the cold war almost everywhere. But it was also a time of great hope and idealism, of rising living standards, of the beginning of the end of colonialism, and of cheap fuel to industrialize the world. Perhaps not since Malthus's time had there been such divergent views about the world's future. Was there, as optimists argued, unending opportunity for freedom and prosperity? Or would the cold war turn hot and trigger nuclear Armageddon? Might there be a world communist revolution? Or would economic hopes be dashed by a Malthusian crunch in the poor world?

By the 1950s, population theory had come a long way. Crude eugenics was put to one side. Knibbs's idea of demographic growth stretching for centuries into the future, until some inevitable Malthusian bust, had been replaced by the idea of the demographic transition. This was the brainchild of Princeton demographer Frank Notestein. In the West, he said, countries had seen their death rates fall as prosperity grew. People responded to the increasing likelihood of their children surviving to adulthood by reducing their birth rate. The transition involved a shift from a world of high mortality and high fertility to one of low mortality and low fertility. Both had stable populations, but in the decades of the transition, when births hugely outnumbered deaths, populations soared.

Notestein devoted his academic life to studying fertility but remained childless. He told the American Philosophical Society in 1943 that the developing world would eventually take the same course of demographic transition already taken by the rich world. But he warned that in the process the world faced "unparalleled population growth." And he argued that a new stable world population would be achieved only if people in Asia, Africa, and Latin America became wealthier in the way that most Western societies had. Without modernization, poor countries could get stuck halfway through the transition—in a trap of falling mortality but sky-high fertility. This was troubling.

A sense of impending Malthusian catastrophe was heightened by fears of a resource and environmental crisis that could engulf rich and poor alike. Knibbs had figured that we had a couple of centuries before we became too numerous for the planet to feed us. But suddenly the threat seemed much closer. One who believed that an environmental crisis was imminent was Fairfield Osborn Jr., the son of Henry Osborn, who had set out to save the redwoods of California, and cousin of Frederick Osborn, the eugenicist. In 1948 he published *Our Plundered Planet*, in which he warned, "There are no fresh lands anywhere. Never before in man's history has this been the case." The only hope, he said, was that "man must recognise the necessity of cooperating with nature. The time for defiance is at an end." His book was well received. But its green message—a forerunner of the modern idea of sustainable development—lost out to a rival publication with a more apocalyptic vision.

·   ·   ·

The father of modern environmental doomsters is all but forgotten today. He merits just a two-sentence stub on Wikipedia. William Vogt was a forty-six-year-old ornithologist with a modest but blameless career editing books on birds at the Audubon Society in New York when, in 1948, he upstaged Osborn with his own book, *Road to Survival*.

It's a real page-turner, even today. Not least because Vogt brought the threats to future human survival disconcertingly close to home. This was no faraway crisis of development. Vogt said the rich would go hungry too, as population rose from just over two billion to three billion by 2050. "It is probable," he wrote, "that at least three-quarters of the human race will be wiped out." We "may yet escape the crash of our civilization," he conceded, but in the tropics, "large areas of the Earth now occupied by backward populations will have to be written off."

There was, he said, already only one acre of cultivatable land for every inhabitant of the planet. And modern technology, while providing more food now, was jeopardizing the future. He singled out the axe, fire, the plow, and firearms, backed up by "the capitalistic system under which the land is managed on the basis of so-called economic laws and in very general disregard of the physical and biological laws to which it is subject." Even gloomy Malthus had regarded the middle classes and the rich as immune from his apocalyptic natural law. For Vogt, the laws of Malthus "touch the life of every man, woman and child on the face of the globe every minute of every day."

He argued that since Malthus, the world had been living on borrowed time, thanks to the colonization of the Americas. The New World had delivered not just new land, but also new crops, like the potato, which had greatly increased the number of people that European fields could feed. Now, however, the game was up.

Vogt singled out Britain. Britannia may have invented the Industrial Revolution and for a time ruled the waves. Now "she was a contented parasite, drawing on the eroding hillsides of New England, of Iowa, of Maryland, of Argentina and South Africa, of Australia and India" to feed herself. Except for coal and Scotch whisky, she had little to offer the world. Vogt held special contempt for "that magnificent anachronism Winston Churchill," who had recently declared, "The earth is a generous mother; she will provide in plentiful abundance for all her children, if they will but cultivate her soil in justice and peace." Far from it, Vogt said. Britain's bloated population was "on the verge of starvation. . . . Unless we [Americans] are willing to place fifty million British feet beneath our dining-room table, we may well see famine once more stalking the streets of London. And hand in hand with famine will walk the shade of that clear-sighted English clergyman, Thomas Robert Malthus."

Britain would be only the first. Industrialization had "reached a point of saturation. . . . The purchasing power of the middle classes, especially the professional groups, is being more and more restricted to the necessities of life. . . . A fall in living standards is unavoidable." Japan "cannot possibly feed itself." The Chinese were "becoming more wretched with every generation." Even U.S. standards of living were certain to decline. Writing little more than a decade after the great Dust Bowl of the 1930s had forced hundreds of thousands of Americans to flee the Midwest, he argued that "American civilization, founded on nine inches of topsoil, has now lost one-third of this soil. . . . We had better enjoy our steaks now, since there will be many less of them within the lifetime of most Americans."

What should be done? What was Vogt's "road to survival"? Here he became yet more strident. "The world is now full," said Vogt. "And the regulation of population has become one of the most essential of international problems." "Death control," by which he meant medical advances, had gone too far. Efforts to further improve health in the developing world were misguided. "The modern medical profession . . . continues to believe it has a duty to keep alive as many people as possible. Through medical care and improved sanitation, they are responsible for more millions living more years in increasing misery. They set the stage for disaster. . . . The greatest tragedy China could suffer at the present time would be a reduction in her death rate."

Vogt set his face against aid for agricultural development. "Building irrigation works, providing means of food storage, and importing food during periods of starvation" had in India simply allowed Indians to carry on in "their accustomed way, breeding with the irresponsibility of codfish." Vogt was writing as the British were leaving India. Massacres were taking place as the imperial lands were divided between Hindu India and Muslim Pakistan. The ornithologist went on, "Steeped in superstition, ignorance, poverty and disease, Mother India is the victim of her own awful fecundity. In all the world there is probably no region of greater misery, and almost certainly none with less hope." With the Brits gone, the turmoil was likely to stultify further economic development. This, Vogt thought, would be a good thing.

Like Malthus opposing parish charity, Vogt hated emergency food aid, even for the starving. The United Nations "should not ship food to keep alive ten million Indians and Chinese this year, so that fifty million may die five years hence." America should stop food exports and leave the "surplus" populations of other countries to be reduced by famine, disease, and, he suspected, war. Future food aid should be predicated on acceptance of strict population controls. In one way Vogt differed from Malthus. For Malthus, contraception was a vice. For Vogt, it was a moral imperative. But you can hear the rumble of eugenic thinking in his call for "the world's shiftless" to be paid to be sterilized, a policy which "would probably have a favorable selective influence."

These were not the meanderings of a maverick. Vogt was the wordsmith for a view that would dominate thinking on the environment and population for half a century, and increasingly influence U.S. national policy, especially on population. A global crisis was at hand in which overbreeders were set to bring down not just the genetic stock of the superior races but also the ability of the entire human race to feed itself.

CHAPTER 5

The Contraceptive Cavalry

The crusade to lower fertility rates in the developing world began in Japan. Until the mid-nineteenth century, when Commodore Matthew Perry of the American navy burst open the door, Japan was a closed and feudal society. But after the fall of shogun rule, Western ideas about health and sanitation began to infiltrate Japan and death rates fell. The country's population doubled to sixty million by the early 1920s and kept on rising. Japan was growing short of land and natural resources. But it had plenty of young men, and its generals sent them overseas to grab more territory on the Asian mainland and in the Pacific. This culminated in the bombing of American ships in Pearl Harbor in 1941 and Japan's entry into the Second World War.

The bid for an empire to sustain its growing population ended in defeat and nuclear humiliation in 1945. Japan was left militarily, politically, and economically crippled—less wealthy than Mexico or the Philippines, and with eighty million people to feed. Disaster loomed. Julian Huxley, the Malthusian biologist now in charge of UNESCO, called Japan "the most overpopulated country there has ever been." William Vogt, in *Road to Survival*, agreed that "she will probably have to face famine. If we attempt to raise her to a standard of living that makes democracy possible, she will remain a world threat—unless her population is systematically reduced."

Enter Princeton's Frank Notestein, the high priest of American demography and theorist of the demographic transition. In 1948 he returned from a visit to Japan believing disaster could be prevented. Japan may be poor, he agreed, but it had been richer before and it might be just the place to try out his latest idea—kick-starting a demographic transition by flooding a country with contraceptives. He had become "convinced that the unprecedented process of population control prior to material well-being is more likely to occur here than anywhere on earth. . . . It might well set the precedent for the whole of East Asia." And as it turned out, he was right.

Japan already had a nascent birth control movement. It had been pioneered by Shizue Ishimoto, the daughter of a traditional samurai family. She became known as "Japan's Margaret Sanger," after the globe-trotting American met her in Tokyo in 1922. Ishimoto's birth control clinics had been closed when Japan's militaristic governments of the 1930s banned contraceptives and promoted population growth for the greater glory of the emperor and Japan. But with postwar Japan humbled and the economy in freefall, government opposition had evaporated.

After a brief baby boom following the return of troops, there was a huge appetite among ordinary Japanese to stop babies. They started with abortions. At one point there were a million a year—many more than live births. But in 1949, America's talismanic birth control entrepreneur, Clarence Gamble, followed Sanger to Tokyo. He funded research that found that 92 percent of the Japanese wanted contraception, and within a couple of years the Washington-controlled government was handing out free condoms in every health center in the land. By 1957, Japanese fertility had crashed from 4.5 babies per woman to 2.0, and then carried on down to 1.3, where it has stuck. Back then, no country had seen anything like it. Notestein's success in voluntary birth control became the start of a movement that would sweep the globe.

· · ·

The terrifying predictions of global famine from Vogt and others clearly required action. The prewar fears of a wave of poor people from the poor world swamping the West were now based on environmental concerns rather than eugenics. And some of the most influential figures in Washington were prepared to take up the challenge. "Between the end of the 1940s and 1952, something quite dramatic happened to the West's attitude to population," says Australian demographer Jack Caldwell in his history of the population movement.

These years saw the independence of a number of former colonies, including India, Pakistan, the Philippines, Indonesia, and Egypt. In the place of European empire was a new world order, based on the United Nations and institutions like the World Bank. Bankrolling it all was Uncle Sam. The United States' Marshall Plan for European postwar reconstruction set a pattern for U.S. engagement in distant countries. And control of soaring population in the developing world was rising to the top of the political agenda in Washington.

If one event put it there, it was a private meeting held in June 1952 in Williamsburg, Virginia, under the auspices of the National Academy of Sciences. Caldwell says it was a "crisis conference," though it was by invitation only and produced no public outcome—not even in the specialist journals. The convener was John D. Rockefeller III, the forty-six-year-old heir to the Rockefeller family fortune that was built on Standard Oil. He was an American philanthropist with deeper pockets than any other. Next to him sat Vogt, whose fame from writing *Road to Survival* had won him the directorship of Margaret Sanger's Planned Parenthood Federation of America. Also at the table were America's two preeminent demographers, Frank Notestein and his former protégé, Kingsley Davis of Texas, along with both the young Osborns, Fairfield and Frederick, the latter an old friend of Rockefeller's.

These were not men troubled by modesty. They saw clearly their right to pronounce on the world's future. Frederick Osborn said they were part of a tiny intellectual elite "of three or four hundred people who produce most of the freedoms of the human mind [and who] may be engulfed by a great mass of people for whom these conceptions are largely alien." They saw the population problem "not as a theory, but as a nightmare." According to notes of the meeting unearthed by historian Matthew Connelly in the Rockefeller archive, both Vogt and Fairfield Osborn argued that "industrial development should be withheld" from countries like India until they adopted population policies.

The three-day meeting decided to set up a Population Council to devise and bring about a global plan to control world population growth. Rockefeller, who personally paid the council members' salaries, was its first president. Frederick Osborn became the first executive vice president, with day-to-day control. The council was soon busy taking in funds from the Ford and Rockefeller foundations. The two postwar giants of private American philanthropy set up laboratory research into developing new and improved forms of contraception and helping countries set up family planning programs, starting in India and Pakistan.

The Population Council was to become the most influential clearinghouse for ideas and policies for family planning and population control. According to its Web site today, that opening meeting heard from John D. Rockefeller that the world had to tackle growing populations "to improve the quality of people's lives, to help make it possible for individuals everywhere to develop their full potential." That may reflect the council's instincts today. But back then there was another, more hard-edged and Malthusian agenda.

Kingsley Davis had told the Williamsburg meeting that food aid was worse than useless. Aid would "build up ever larger populations on the basis of charity" and it "would become ever more difficult to remove the prop." He also opposed migration, particularly the United States' habit of "acquiring each year tens of thousands of impoverished, illiterate, superstitious, non-English-speaking, and in many cases diseased, new citizens." This at a time when migration into the United States was at a historic low.

Trade with poor breeders was also deemed a bad idea. Vogt had said in 1948, "Why the United States should subsidize the unchecked spawning of India and other countries by purchasing their goods is difficult to see." The old prewar eugenic thinking also remained. The desire to cut population growth was surprisingly selective. Frederick Osborn, who resigned as president of the American Eugenics Society to take up his post at the Population Council, dropped into the first draft of the council's mission statement an aim to create conditions under which "parents who are above the average in intelligence, quality of personality and affection will tend to have larger than average families." The sentence was dropped, but he still saw no inconsistency between his new role as the council's CEO and writing in 1956 in the *Eugenics Review* that "we need the greatest number of births among genetically superior individuals."

. . .

Liberals may find much of this unpleasant today. But there is no disguising the gravity of what these would-be masters of demography were tackling. Medical and public health technology was bringing down death rates so far and so fast in some poor countries that their populations were experiencing growth much more sudden than anything seen in Europe or North America during this phase of their demographic transitions.

At UNESCO, Huxley noted in 1955 that "in England, malaria took three centuries to disappear; in Ceylon [modern Sri Lanka] it was virtually wiped out in less than half a decade, thanks to DDT." As a result, the island's death rate was halved in seven years, and life expectancy rose from forty-seven to sixty years. It was the statistical equivalent of everlasting life: every year you lived, you could expect to live two more. The population growth rate of 2.7 percent was twice the highest rate experienced in industrializing England.

Of course, according to Notestein's transition theory, this population growth should not go on forever. Fertility would fall. But Notestein had concluded that industrialization and urbanization, plus the spread of wealth, were essential before that happened. Thanks to the introduction of Western death control measures, countries like Sri Lanka and dozens of others across the developing world had entered the first phase of the transition far earlier than they otherwise would. And it was not clear whether or when modernization would allow fertility to start falling. These countries could get stuck for decades, or even centuries, with low mortality but high fertility. So following their success in Japan, the population controllers decided to turn the demographic transition from description into prescription. Rather than seeing it as a process that scientists could plot but not influence, they decided to kick-start a decline in fertility by intervening directly to change the reproductive habits of poor people in poor countries.

Since his visit to Japan, Notestein had argued that falling fertility need not simply be a consequence of modernization; it could in fact drive the process. Bringing down fertility should make people richer, stimulating a further fall in fertility, and so on. His colleagues at Princeton bolstered the case. A study by demographer Ansley Coale and economist Edgar M. Hoover in 1958 argued that bringing up lots of children stopped the poor from investing their energies in wealth creation. They reasoned that cutting out the babies could increase investment and make people richer quicker.

Many developing countries rejected this argument. They wanted cash for factories and fertilizer, not shiploads of condoms. "Economic development is the best contraceptive," they said. But the new population czars from Washington were not going to wait on economic development. And it is easy to see why. Postwar Indonesia was on course to double its population in thirty-one years, Nigeria in twenty-eight years, Turkey and Kenya in twenty-four years, Brazil in twenty-two years, the Philippines in twenty years, El Salvador in nineteen years, and Costa Rica in seventeen years. And far from being a one-generation spurt, the increase in fertility rates in many of these countries was continuing. The average woman in the developing world was having almost six children. Births outnumbered deaths two or three to one. In 1950 the world's population stood at just under 2.6 billion, and the United Nations' estimate for 1980 was 3.3 billion. A decade later, population has already hit 3 billion, and the prediction for 1980 had risen to 4.2 billion.

For a while, it wasn't just poor countries that were having a baby boom. In Europe and North America, soldiers returning from war had, unsurprisingly, celebrated peace by starting families. But the boom did not stop. American women in 1957 were having an average of 3.8 children—a figure not seen since the 1890s. Newfound wealth, and homes stuffed with labor-saving devices like vacuum cleaners and washing machines, encouraged larger families. And with their menfolk reaping riches in the postwar economic boom, women were still mostly staying at home after marriage.

But the concern in Washington was less with breeding American moms than with events in far-off lands. As the 1950s progressed, the response to the population crisis was increasingly shaped by the cold war between capitalism and communism.

One insider with a flair for plain speaking was millionaire Hugh Moore, the founder of the Dixie Cup Corporation, which introduced the original watercooler and its ubiquitous Dixie cups. Moore had become a big hitter in cold-war Washington, as treasurer of the Marshall Plan and the Woodrow Wilson Foundation and president of an organization called Americans United for World Organization. He said he had experienced "a religious revelation" on reading Vogt's *Road to Survival*. And in 1954 he circulated millions of copies of a pamphlet he had written called *The Population Bomb*—fourteen years before Paul

Ehrlich's book of the same name—and devoted the rest of his life to curbing population growth. In a note to Rockefeller introducing his 1954 pamphlet, he declared, "We are not primarily interested in the sociological or humanitarian aspects of birth control. We are interested in the use which Communists make of hungry people in their drive to conquer the earth."

Such talk set alarm bells ringing in the Pentagon. In 1958 a high-powered committee was set up by President Dwight Eisenhower to "appraise the military assistance program." The committee was chaired by General William H. Draper, a former U.S. ambassador to NATO and before that one of the leaders of the American eugenics movement. He was also an old friend of Moore, who deluged the committee with papers giving his views on the population problem. Draper's report concluded that hungry people without enough land to grow food were likely to be seduced by dreams of land reform. That was socialism. So the "population explosion" was a strategic threat to the United States.

For the moment, the crusade to stop babies was mostly in the hands of wealthy and influential enthusiasts like Rockefeller, Moore, and their associates at the Population Council—social entrepreneurs, we'd call them today. Outside the United States, a major force was the International Planned Parenthood Federation (IPPF), formed in 1952 at a conference in India. Its founding members were the privately funded family planning associations of Britain, India, the Netherlands, Hong Kong, Singapore, Sweden, West Germany, and the United States.

Today the IPPF's Web site says its formation was "the result of campaigning by a handful of brave and angry women, including Elise Ottesen-Jensen from Sweden and Dhanvanthi Rama Rau from India, who were imprisoned for their assertion that women had the right to control their own fertility." True. But like the modern gloss from the Population Council, it hides another agenda. From the start, the IPPF was an alliance of the old eugenics lobby and a new generation of women more concerned with reproductive rights. Other original board members now largely airbrushed from the story include its first president, the controversial eugenicist Margaret Sanger, and a future chairman, Carlos Blacker. An Eton-educated psychiatrist and acolyte of Julian Huxley, Blacker had for more than two decades been secretary of the British Eugenics Society—which hosted the IPPF in its London offices.

By the early 1960s, this private network had dozens of national associations distributing contraception. Most of the money came from the United States. Tens of millions of dollars from the Ford and Rockefeller foundations and General Draper's newly established Victor-Bostrom Fund passed through the Population Council and Clarence Gamble's Pathfinder Fund, which was now at work in more than thirty countries. Gamble led the charge in Bangladesh—where he eventually spent $200 million on family planning services—as well as in key battlegrounds of the cold war, like Indonesia, Egypt, and much of Latin America.

The network was also funding the development of contraceptive technology. Until the 1960s, there was the condom, the fiddly diaphragm, withdrawal, some frankly dangerous IUDs (intra-uterine device), the rhythm method—and abortion. But the booming interest in population control led to the development of a smorgasbord of new and improved methods, many of them pioneered at the Population Fund's labs and field-tested at the expense of Clarence Gamble. At the start of the decade, researchers developed a new IUD, the Lippes Loop, to replace two old IUDs that often dragged bacteria into the uterus and caused pain, bleeding, and pelvic inflammatory disease. The new IUD was followed by hormonal contraceptives—first the pill and later injectables such as Depo Provera. The revolution was completed by better methods of sterilization for both sexes (particularly tubal ligation for women) and safer techniques of abortion, using suction. This was mostly good work. But some of the motivation and practical application was dubious, to say the least.

An early test bed of the United States' private population conquistadors was Puerto Rico. The Caribbean island had been under U.S. control since 1898. It was first labeled "overpopulated" as early as the 1930s. In 1948 Vogt said of Puerto Rico, "The population has doubled during the years since it became an American colony. For each inhabitant there remains less than one half-hectare of arable land, and much of the land remaining belongs to continental sugar corporations which export its products for their own profit." As a result, thousands of landless farmers headed for the capital, San Juan. One solution would have been land reform—to take back some of those sugar plantations for the poor peasant farmers. Vogt did not mention that possibility. Instead he blamed the peasants. "Puerto Rico is poor in resources and almost without power—

except the power to reproduce recklessly and irresponsibly." What was required was not land reform but control of the fertility of these feckless farmers.

America had first tried out sterilization in Puerto Rico in the early 1930s, at the Presbyterian Hospital in San Juan. It was an early foreign venture for Gamble, who at that time had an avowedly eugenic agenda. Sterilization was chosen because, according to modern U.S. demographer Harriet Presser of the University of Maryland, "many physicians felt that low-educated Puerto Ricans were generally ineffectual users of contraceptive methods." This work continued into the 1940s and 1950s, when Gamble was among U.S. philanthropists who established the Family Planning Association of Puerto Rico (Profamilia). By the 1960s, the program had sterilized a third of all Puerto Rican women of childbearing age.

Profamilia also began funding field trials in Puerto Rico of IUDs and then the pill. The pill was invented by American gynecologist John Rock and hormonal biologist Gregory Pincus, after Margaret Sanger challenged Pincus to come up with a "magic pill" to prevent conception. After preliminary studies in the United States, Rock and Pincus needed a place to conduct large-scale trials of the pill's effectiveness, safety, and suitability among poor, uneducated women. They quickly settled on Puerto Rico, which already had a network of birth control clinics and a large client base.

The drug company G. D. Searle provided the high-dose progesterone pills, and Gamble again paid the bills. The trial was a success —except that 17 percent of the women reported symptoms like nausea, dizziness, stomach pain, and vomiting from the high progesterone doses in the pill. The American doctor running the trials, Edris Rice-Wray Carson, asked for lower doses. Pincus ignored her, and he carried out no studies into side effects. It also turned out that the women were never told they were taking part in clinical trials. But the conquistadors were happy. The combined force of widespread sterilization, IUDs, and the pill halved the island's fertility rate to 2.7 babies per woman by the mid-1970s.

· · ·

In the 1950s and early 1960s, governments had mostly been happy to take a backseat to private family planning philanthropists. Colonial powers feared being accused of what one British official in Hong Kong called "a desire by a white government to restrict growth of the non-white population." Eugenics, in other words. However, the privateers of contraception were determined that their cause should be taken up by the U.S. government. Having a Catholic president, John F. Kennedy, was tricky for them. But when his place was taken by Lyndon Johnson, their message gained traction. In 1965 Johnson told Congress, "I will seek new ways to use our knowledge to help deal with the explosion in world population and the growing scarcity of world resources." The U.S. government's Agency for International Development (USAID) opened a population office and announced that it was now supplying contraceptives. With Johnson paying, family planning programs spread. Turkey, Malaysia, Egypt, Chile, Morocco, Kenya, Jamaica, Iran, Colombia, Costa Rica, Taiwan, South Korea, and Tunisia all swiftly came on board.

By the late 1960s, the U.S. government was rolling out family planning round the world, spending $100 million a year and for the first time outspending the private U.S. foundations. There were new U.S.-funded government family planning programs in Hong Kong, Singapore, Sri Lanka, and Brazil. In a few places the privateers survived. In Thailand, family planning was pioneered by a maverick politician known as "Mr. Condom King," Mechai Viravaidya. He handed out condoms free at his restaurant chain, Cabbages and Condoms. Condoms are sometimes know there today as Mechais.

Back home, the U.S. pioneers used their influence with government to strut the stage. General Draper established the Population Crisis Committee (later Population Action International) as his personal mouthpiece and went to Congress to demand more cash. He testified to the House of Representatives that "unless and until the population explosion now erupting in Asia, Africa and Latin America is brought under control, our entire aid program is doomed to failure." From there, he moved on to represent the United States on population at the United Nations. At his instigation, the United States bankrolled the establishment in 1967 of the UN Population Trust Fund, which two years later became the UN Fund for Population Activities (UNFPA). This was now truly global business.

In 1966 Rockefeller organized a joint declaration by thirty world leaders that a "great problem threatens the world . . . the problem of unplanned population growth." The declaration said that access to family planning methods was "a basic human right" that enabled "enrichment of human life, not its restriction." Its signatories included Harold Wilson in London and Ferdinand Marcos in Manila, Gamal Abdul Nasser in Egypt and Indira Gandhi in India, King Hussein of Jordan and Marshal Tito, the strong man of Yugoslavia. In 1969 Rockefeller got his nominee—a Filipino civil servant called Rafael Salas, who had run the country's successful national rice sufficiency program—appointed to head the UNFPA.

The final piece fell into place in 1968, when Johnson nominated his defense secretary Robert McNamara as president of the World Bank. McNamara swiftly declared that in the future, aid for health care would be tied to population control. Death control, he said, was not acceptable without birth control. "To put it simply," he said, "the greatest single obstacle to the economic and social advancement of the majority of people in the underdeveloped world is rampant population growth." The threat was "very much like the threat of nuclear war."

Countries that held out against the doctrine came under pressure. In 1973, after visits from McNamara and Draper, Mexico reversed its previous support for population growth and embraced a family planning program. And where pressure didn't work, USAID bypassed governments and threw its money into the voluntary sector via Clarence Gamble.

. . .

There was one problem. So far as anyone knew, all this effort was having little impact. Right through the 1960s, the world's population continued to grow by more than 2 percent a year. Each year there were seventy million more mouths to feed. Some began to recall the words of Kingsley Davis, who in 1967 had written in *Scientific American*, "The things that make family planning acceptable are the very things that make it ineffective for population control. By stressing the right of parents to have the number of children they want, it evades the basic question of population policy, which is how to give societies the number of children they need."

Was compulsion required? Maybe mass medication would be in order? An internal report at the Ford Foundation suggested research into "an annual application of a contraceptive aerial mist (from a single plane over India) neutralized only by an annual antidotal pill on medical prescription." The suggestion didn't get very far. But it is revealing that such an idea could even get past the front door.

CHAPTER 6

Three Wise Men

Paul Ehrlich was a biologist in California. He spent most of his early professional life watching butterflies. He had noted how, given the right conditions, they would breed so fast that they consumed all the plant leaves they could eat. Having exceeded the "carrying capacity" of their immediate environment, they became extinct. It rang a bell. Ehrlich had read William Vogt as a teenager. Humans were now, he concluded, like the marauding butterflies. We had exceeded the carrying capacity of the planet. We could no longer be fed on the land available. It was as simple as that.

Ehrlich lived the message. He had a vasectomy in 1963, after he and his fellow-scientist wife Anne had one child. In 1967 he wrote an article on the problem for *New Scientist* magazine in London, later reproduced in the *Washington Post*. David Brower, the director of Friends of the Earth in the United States, read the piece and commissioned a book. A month later he had the manuscript for *The Population Bomb*, a title that revived Hugh Moore's old slogan but for a mass audience. *The Population Bomb* eventually sold some six million copies and made Ehrlich a media star after he appeared on Johnny Carson's *The Tonight Show*. Demography was show biz. Population was out of the seminar rooms and onto the streets.

Ehrlich predicted that the 1970s would see "the greatest baby boom of all time." Some 40 percent of the population of the "undeveloped world" was under fifteen years old, with their whole period of procreation ahead of them. So the baby boom was pretty inevitable. But

Ehrlich's central claim was that there would be no food for this new generation. "The battle to feed humanity is over. . . . Billions will die in the 1980s."

In many ways Ehrlich did little other than to repeat what Vogt had argued two decades before. But his message was even more urgent, because population was growing far faster than Vogt had ever imagined. By 1968, it had already exceeded the three billion figure that Vogt had warned could be reached by 2050, and was set to double again in thirty years or less, not a century. I was a school student then. I remember wondering how the world could feed a population set to double in a generation. It didn't seem possible.

In *The Population Bomb*, Ehrlich offered several possible futures. At worst, a billion people (almost a quarter of the world's population at the time) would starve to death by 1983. At best, thanks to a crusade by the pope in favor of birth control, the billionth death would be delayed until 1990. However it panned out, "nothing could be more misleading to our children than our present affluent society. They will inherit a totally different world, a world in which the standards, politics and economics of the past are dead." For a sixteen-year-old would-be hippie, this was scary stuff, man.

The scientific origins of this apocalyptic view may have been butterflies, but the emotional beginning was even more interesting. It started, Ehrlich said in the book's first chapter, in a late-night taxi ride through Delhi with his wife and daughter. "The streets seemed alive with people. People eating, people washing, people sleeping, people visiting, arguing and screaming. People thrusting their hands through the taxi window begging. People defecating and urinating. People clinging to buses. People herding animals. People, people, people, people. As we moved through the mob, hand horn squealing, the dust, noise, heat and cooking fires gave the scene a hellish aspect. Would we ever get to our hotel? All three of us were, frankly, frightened. . . . Since that night I've known the *feel* of overpopulation."

It is a racily written passage. But it's revealing, as Jack Caldwell, the Australian demographer whose specialty is people rather than butterflies, later pointed out. Ehrlich "did not see population explosion, for Delhi's birth rate is relatively slow. He probably saw fewer people than one would see with pleasure in New York, London or Paris at Christmas or in the peak hour. What he did see were poor non-Europeans." Ehrlich's rhetoric also prompted Caldwell to remember how "the eugenics movement exhibited aspects ominously similar to some facets of the contemporary population and conservation movements."

Despite declaring that billions of us were doomed, Ehrlich had a series of policy prescriptions that suggested otherwise. He said, for instance, that there was "no rational choice except to cut off food aid unless areas carried out strict and compulsory population control programs." There would also have to be controls on migration to prevent those without aid from moving to areas with aid. He advocated establishing an optimum population for the world. "At a minimum it seems safe to say that a population of one or even two billion people could be sustained in reasonable comfort.... At four or five billion we will still have a chance." And he wanted a detailed program of contraception to limit population "by compulsion if voluntary methods fail."

His views were widely held. Harriet Presser, who trained as a demographer under Kingsley Davis in the 1960s, remembers, "The talk then was that there should be a license to have children, the same as you needed a license to have a dog."

. . .

Ehrlich's book was the public potboiler for the population control movement. But of equal importance among academics was the work of a rather older California biologist. Garrett Hardin had for many decades been professor of biology at Santa Barbara. He was a man of principle. Rather like the young Malthus, he insisted on going where his logic took him, regardless of who objected. He was an outspoken eugenicist. In 1951 he had argued that "every time a philanthropist sets up a foundation to look for a cure for a certain disease, he thereby threatens humans eugenically." Doctors were undermining the natural law of the survival of the fittest. Hardin held this uncompromising view throughout his life.

It reached its height in 1974, when he wrote a paper called "Lifeboat Ethics" that argued that in the modern world "each rich nation can be seen as a lifeboat. In the ocean outside swim the poor people of the world, who would like to get in." But there were not enough resources to go round. The people in the lifeboat had a duty to their species to be selfish—to keep the poor out of the lifeboat, even if they drowned as a result. That meant no food aid, no migration to rich countries, no medical aid. He quoted with approval Alan Gregg, a former vice president of the Rockefeller Foundation, who in the mid-1950s had described humanity as a cancer, adding that "cancerous growths demand food, but as far as I know, they have never been cured by getting it."

Hardin argued that "every Indian life saved through medical or nutritional assistance from abroad diminishes the quality of life for those who remain, and for subsequent generations." Aid should be given only "if they are taking aggressive measures to reduce population." Our species' survival, he concluded, "demands that we govern our actions by the ethics of the lifeboat, harsh though they may be."

Many greens can't quite stomach that. They prefer instead Hardin's eye-catching paper in *Science* in 1968, called "The Tragedy of the Commons." It argued that when people share a common resource—whether of land or clean water or anything else—those who take (or pollute) the most gain the most, while everybody suffers the consequence of the resource's decline. The classic case is animal herders using a common pasture. Those with the most animals will make the most profit. But everyone, rich and poor, will eventually suffer as the pasture is overgrazed. Soon after Hardin wrote this paper, in the early 1970s, environmentalists blamed the desertification of the Sahel in Africa and the subsequent famine on just this process. Today we see another global commons under threat: the atmosphere. We are polluting the atmosphere, causing a buildup of greenhouse gases that are destabilizing climates. Everybody will suffer in the end. But meanwhile it is those who are causing the most pollution who get rich.

Hardin said that in order to survive, we had to abandon the commonality of the commons. He did not stipulate that this amounted to privatization. He accepted that ancient societies had other, communal ways of managing the commons. But he saw that as socialism and beyond the pale.

Meanwhile, he said, more commons would have to go. "In a more embryonic state is our recognition of the evils of the commons of pleasure." He buttered up his readers with some familiar gripes: Muzak in shopping malls, intrusive advertising and aircraft noise. Then came the rub: "We must now recognize the necessity of abandoning the commons in breeding. Freedom to breed will bring ruin for all." It must be "relinquished to preserve and nurture other and more precious freedoms."

Many people favored voluntary birth control, he said. The Universal Declaration of Human Rights held that "any choice and decision with regard to the size of the family must irrevocably rest with the family itself." But he argued that this ethical taboo "must be resisted, because an appeal to independently acting consciences selects for the disappearance of all conscience in the long run." Conscience-free overbreeders would take over the planet by sheer force of numbers. Eugenics again. The "only way" to save humanity was through a "fundamental extension of morality" to "relinquish the freedom to breed, and that very soon." No wonder the Web site extolling his work is called "Stalking the Wild Taboo."

Hardin's law of the commons has proved as popular as Malthus's law. *Science* magazine has received more requests to reprint this paper than any other in its history. But it is also reviled. American anthropologist Eric Ross says it "embodies all of the cardinal qualities of cold war Malthusian thinking: it is anti-socialist, anti-democratic and eugenic." Many environmentalists revere Hardin. I find most of his ideas abhorrent. But I'll say this for him: he said what he meant.

. . .

While Ehrlich hit you in the gut and Hardin grabbed your intellect by the throat, another study played out the future of the planet in a way that we might now see as a computer game. The year 1972 saw the publication of a slim volume called *The Limits to Growth*, authored mostly by Donella Meadows and displaying some graphs produced by her husband, Dennis Meadows, a systems analyst from the Massachusetts Institute of Technology. It was published not as a peer-reviewed paper, but rather as a book released at a huge party in the gothic Great Hall of the Smithsonian Institution in Washington. The party was paid for by the Club of Rome, a private institute set up by Aurelio Peccei, an Italian millionaire industrialist and wartime antifascist hero, who sent twelve thousand copies of the slim volume to ministers and journalists round the world.

Meadows computed the future with a model of the world economy conceived by MIT professor Jay Forrester. It contained just five main parameters: population, food, industrial production, pollution, and scarcity of nonrenewable resources like metals. But it produced some impressive-looking graphs suggesting that unless current trends in population growth and industrial output were checked, and pollution staunched, civilization faced collapse within a century, perhaps within fifty years.

The standard scenario showed business continuing as usual till about 2010. After that, "as resource prices rise and mines are depleted, more and more capital must be used for obtaining resources, leaving less to be invested for future growth. Finally investment cannot keep up . . . and the industrial base collapses, taking with it service and agricultural systems. . . . Population finally decreases [around 2050] when the death rate is driven upwards by lack of food and health services."

Other runs of the model assumed we wouldn't go like lambs to the slaughter. One figured we would try technical fixes like recycling materials, clamping down on pollution, and offering free family planning. Another thought we'd all become less consumerist and each have no more than two children. Even these scenarios eventually delivered "overshoot and collapse."

Meadows insisted that this was simply a "sensitivity analysis," aimed at showing what the critical tipping points might be in a future world of environmental limits. But that was the small print. This was a time of computer innocence. As the British magazine *New Internationalist* remarked, "Computer portents of doom have a power to impress and terrorize which the ordinary mortal does not possess. The cool reasonability and the inevitability which attends the hundreds of millions of microcircuits, seems to banish all possibility of argument." Caldwell, no doubt miffed at a systems analyst invading his patch, agreed: "Forrester's changing curves across the terminal screen looked more conclusive before the age when any child could alter such lines at will on the family TV screen with some games software."

The Limits to Growth certainly grabbed attention. *Science*, the voice of American science, ran five pages on it. The article noted that "the book reveals none of the assumptions and equations that are the meat of the model." When these were finally published, critics said the apocalyptic conclusions had been fixed from the start. The formulas put into the model were Malthusian to the core. All the bad things—population, pollution, our demand on resources—were set to increase exponentially, while all the good things, like technological breakthroughs, increased only arithmetically. Surprise, surprise, the world sank into a mire of pollution, soaring commodity prices, and famine. Whatever its shortcomings, the report had found its moment. Within two years, the OPEC oil shock and soaring prices for key commodities made it look like the world was fast-forwarding into Meadows's future.

The evidence seemed to be piling up. Something had to be done fast about population. With China, the world's most populous country, an island unto itself, neither demanding nor receiving any assistance, attention concentrated on the world's second most populous nation: India. Its soar-away population growth, everyone agreed, had to be halted.

Six Dollars a Snip

There is a long tradition of Malthusian doomsaying about India. In the early nineteenth century, virtually every British colonial satrap had heard Malthus lecture at the East India College before he set foot in the country. Just like the British administrators in Ireland, they blamed the country's repeated famines on overpopulation, even as shiploads of grain left Indian ports bound for Europe. So it was in the late 1870s, when between five million and eight million people starved to death under the Raj, including a quarter of the population of Mysore. The colonial administrator Sir Evelyn Baring told Parliament in London: "Every benevolent attempt to mitigate the effects of famine . . . serves but to enhance the evils resulting from over-population."

By the mid-twentieth century this complacent fatalism was complemented, if not yet replaced, by growing interest in birth control to check population. An inquiry into the Bengal famine of 1943, in which a further three million people died, recommended setting up birth control clinics. Britain was on the verge of abandoning India, but India's indigenous elites had long been keen to defuse its population bomb. The Maharajah of Mysore had opened the world's first government birth control clinics in 1930. Five years later, the All-India Women's Conference invited Margaret Sanger to discuss a national program. The leader of the country's independence movement, Mahatma Gandhi, favored abstinence (he slept with young women in order to test his personal self-control), but birth control was more generally favored among the middle classes.

At independence in 1947, India had a population of about 350 million. But it was rising by almost 2 percent a year as health programs, including a massive effort to eradicate malaria, raised life expectancy at birth from thirty to thirty-six years by the mid-1950s. The fertility rate was also rising as healthier women were having more babies. It reached 6.1 children per woman in the early 1950s. There were twice as many births as deaths. The government forecast that India's population would reach 775 million by the mid-1980s, which turned out to be spot on. But it also forecast that as a result, its people would become 1 percent poorer each year, which turned out to be very wrong.

Jawaharlal Nehru, the country's first prime minister, was hot on population control. "We produce more and more food, but also more and more children. I wish we produced fewer children." Early in 1952, Nehru published the world's first population control plan. It concentrated on encouraging vasectomies. "Our need is desperate, the claims of humanity appeal to us, and it is essential that we should do something for regulating population," said his vice president, Dr. Sarvepalli Radhakrishnan.

But Indians still liked large families. Radhakrishnan knew this well. He had been married at sixteen to a cousin who was just ten, and they had six children. In the first decade of the population control plan, fertility did not fall. Instead it rose again, to 6.5 children per woman. It was at this point that the Americans got involved. And with Lyndon Johnson in the White House, they were talking money.

In 1964 Stephen Enke, an economist from the Rand Corporation, calculated the cash benefit to the Indian economy of sterilizing a man. It was $279, the difference between the cost of feeding the boy for fifteen years till he grew up and the man's likely income during adult life. Actually, the finding was a fix. Enke had set a discount rate in his calculations of a whopping 15 percent per year. This discount meant that for every year that India had to wait for the income generated by the working adult, he lopped off 15 percent of its value. So the future adult's income was rendered virtually worthless in comparison with the cost of feeding the child. But if you forget the small print, Enke's calculation made vasectomies, at four dollars a time, a sound investment for U.S. aid money.

Johnson bought it. Within weeks he told a UN meeting, "Let us act on the fact that less than five dollars invested in population control is worth a hundred dollars invested in economic growth." And soon he had some leverage. In 1966 and 1967, India suffered two successive droughts. In the first year, the United States shipped a fifth of its wheat harvest to India. Johnson wondered at the wisdom of this. Well, he did more than wonder. According to his senior adviser, Joseph Califano, "Johnson exploded. 'Are you out of your fucking mind? I am not going to piss away foreign aid in nations where they refuse to deal with their own population problems.'"

Johnson found an unexpected ally: the newly elected Indian prime minister, Nehru's daughter, Indira Gandhi. As minister for information, she had run aggressive family planning propaganda for her father. Now she wanted to do more than exhort. After the two leaders met in March 1966, Johnson reported back to Congress that "the Indian government believes that there can be no effective solution of the food problem that does not include population control. The choice now is between a comprehensive and humane program for limiting births and the brutal curb that is imposed by famine."

He meant that not as a Malthusian prediction but as a threat: curb births or forget about food aid. And that choice was transmitted intact down to every village in India. At no point did anyone explicitly state that poor people would be left to starve if they did not accept IUDs or sterilization. But in effect, in the midst of a food shortage, that is what happened. Madras started paying its citizens six dollars a snip. Then Kerala, Mysore, and Maharashtra, including Bombay. While average sustenance slipped below a thousand calories a day, the sterilization rates soared as people took the cash. For the poor, this was a famine survival strategy. A report at the Bihar Ministry of Public Health artlessly noted, "The large number of sterilizations during 1967–68 was due to drought conditions." Tribals and Dalits and other poor outcasts were disproportionately operated on. It was the compulsion of an empty stomach.

The end of the famine did not affect Gandhi's resolve. She cannot have been pleased to read Paul Ehrlich's 1968 prognosis that India was "so far behind in the population-food game that there is no hope that any food aid will see them through to self-sufficiency." But undaunted, the government in 1968 set a goal to reduce the Indian birth rate by 45 percent within a decade. Sterilization was industrialized. Two years later,

Kerala organized the first mass vasectomy camp in Ernakulam, north of Kuchin. It lasted a month. There was, by local accounts, a carnival atmosphere, with big incentives for officials and cash payments sometimes equivalent to a month's wages for the fifteen thousand men who accepted the snip. Ninety doctors working in two shifts performed vasectomies in fifty cubicles. Operations began at nine a.m. and continued well into the night. Across the state, doctors carried out sixty thousand vasectomies a month, sufficient to render the state's entire adult male population impotent in a decade. Gujarat went one better, claiming a world record for sterilizing 223,000 men in sixty days. No country had ever attempted to take the scissors to its fertility in such a fashion.

But when the 1971 census returns came in, they revealed that despite this frenzy of activity, the country's population had risen by 109 million since 1961—more than the entire population of all but four other countries. And the sterilization program was faltering. Most of the people willing to be sterilized had already had it done. In 1974–75 just 1.4 million operations were carried out. And many of those had more to do with meeting targets than cutting the birth rate. Attracted by the cash, senile eighty-year-old men were being sterilized in Madras. Over half the men snipped in Uttar Pradesh were over fifty. Women had their new IUDs removed by village midwives so they could go back and have another one, and collect more cash.

Gandhi faced a more immediate crisis. Courts had found her guilty of using the machinery of government to ensure an election victory. In response, she organized the declaration of a state of emergency and began rounding up her opponents. With the constitution suspended, she took the opportunity to impose a new National Population Policy. The gloves were off. The policy stated, "It is clear that simply to wait for education and economic development to bring about a drop in fertility is not a practical solution. The very increase in population makes economic development slow and more difficult of achievement. The time factor is so pressing, and the population growth so formidable, that we have to get out of this vicious circle through direct assault upon this problem. . . . Where [an Indian] state legislature, in the exercise of its own powers, decides that the time is right and it is necessary to pass legislation for compulsory sterilization, it may do so."

The Rubicon had been crossed, the C-word uttered. And the response from the international population agencies was quick. The UNFPA, the London-based IPPF, the Swedish International Cooperation Development Agency (SIDA), and others all increased their funding for Indian family planning. World Bank officials in Delhi cabled home asking for cash to support the "new more vigorous family planning campaign." The bank's boss, Robert McNamara, wrote, "At last, India is moving to effectively address its population problem."

Gandhi, echoing Garrett Hardin, explained, "Some personal rights have to be kept in abeyance . . . for the human rights of the nation, the right to live, the right to progress." She set her youngest, and reputedly rather indulged, son Sanjay the task of pushing through the policy. He was then just twenty-nine years old and held no official position. He combined a revival of sterilization camps with more brutal acts, many organized by his friend Jagmohan of the Delhi Development Authority. And he targeted Muslims.

At three a.m. one November morning, police south of Delhi surrounded the Muslim village of Uttawar. The villagers were woken by loudspeakers ordering men over fifteen years old to assemble at the bus stop. When they obeyed, about four hundred were taken away, mostly to the holy Hindu city of Palwal, where they were forcibly sterilized. The village was singled out because it had protested against the sterilization plans. In Delhi, Sanjay and Jagmohan began the violent "cleansing" of the city's Muslim slums. Around a quarter million people had their homes demolished. Couples had to produce a sterilization certificate if they wanted to apply for new housing. This came only two weeks after police killed more than forty people during demonstrations against compulsory sterilization in Muzaffarnagar in Uttar Pradesh.

All over the country, even lowly government officials received targets for the number of "volunteers" for sterilization that they had to bring to the camps. This was no carnival anymore. The carrot had been replaced by the stick. Sterilization became a condition of receiving everything from rickshaw licenses to medical care, and irrigation water to ration cards. All across the country, state governments began jailing people with three or more children who refused sterilization. During the two years of the emergency, more than nineteen million sterilizations were carried out in India, three-quarters of them on men.

As one Indian journal later reported, "Thousands upon thousands of people, mostly poor and illiterate, were herded like cattle to face the butcher's knife and then to become statistics of targets achieved two and three times over.... Whether the person on whom the knife was wielded was 18 years old or 60, whether he was married or unmarried, whether he had six children or none, became matters of irrelevance since the objective was to tote up awesome figures as proof of loyalty to the powers-that-were in New Delhi."

Were the international funders of all this worried about the compulsion? Seemingly not, whatever the UN Declaration on Human Rights said. They preferred to look at the numbers. At the height of the campaign, McNamara visited the family planning minister Karan Singh, son of the last princely ruler of Kashmir, to congratulate him. Back in Washington, McNamara's staff told the UNFPA they had no formal policy for or against compulsory sterilization. Peter Hegardt of SIDA reported to Stockholm that "even young and unmarried men more or less are dragged to the sterilization premises." That didn't stop his bosses from increasing funding.

It was a common view then that in the "war" against population growth, whatever worked went. A year before, in the first issue of the Population Council's new journal, *Population and Development Review*, Kingsley Davis, the dean of American demography, said that "the only force capable of managing population growth appears to be a strong government." He forecast the need for "a totalitarian government . . . ruling a docile mass of semi-educated but thoroughly indoctrinated urbanites . . . accepting passively what is provided for them."

The coercive policy was a tragic mistake. Thousands died from botched sterilizations during the Indian emergency. And the semi-educated masses proved far from docile. Newspapers were banned from reporting either the deaths or the widespread demonstrations against the sterilization program. So when Gandhi declared an end to the emergency and called a new general election, she must have expected to win easily. But the largest democratic vote in history expelled her Congress Party after thirty years in power. As *Time* magazine later concluded, "One issue, above everything else, cost Indira Gandhi the election: her mass sterilization campaign."

The denials began. Nobody in the international community, it seemed, had sanctioned, colluded in, or even been aware of the contra-

ceptive cudgel employed on the people of India—even though it had
been signaled fairly clearly in Gandhi's National Population Plan. Only
the Australian demographer Jack Caldwell seems to have acted while
the tragedy was unfolding. He recalled in *Science* in 2008 how he and
his wife Pat had "rushed to Delhi to tell scholars and diplomats what
was happening; the latter replied that we should not interfere with Indians doing things in their own way." Sadly, he didn't mention how the
scholars responded.

Many of those who looked the other way at the time now say
the campaign put back the cause of family planning in India by decades. The British demographer John Cleland says it was all largely
unnecessary. "Had Indira Gandhi known the real power of voluntary
family planning—the combination of modern contraception and safe
abortion . . . her government would not have gone down such a sad and
disastrous path." But the same might be said for the entire population
control establishment.

In any event, there was a backlash under way against coercive population control—one that would have far-reaching implications. It had
begun quietly before the Indian emergency. In Washington, the Population Council and the Rockefeller Foundation had begun hearing about
popular protests in some countries against their programs. And even
internally there was unease. A younger generation of activists, especially
women, didn't like some of the attitudes or methods of their male elders.
Rockefeller's first female assistant, a women's health activist named Joan
Dunlop, complained to him in a memo that "a very small number of
men control all the money and the ideas in this field."

As well as being concerned about population, these younger activists were feminists and libertarians, products of the generation of flower
power and Greenpeace and anti–Vietnam War protests. For them, the
cold-war rhetoric and Hardin's lifeboat ethics were anathema. They
turned up in force at the UN environment conference held in Stockholm in June 1972, where they challenged the idea that the planet's
greatest ecological problem was too many poor people. Rather, it was
overconsumption on the part of the rich world. They also challenged
the old guard at a World Population Conference held in 1974 in Bu-
charest. Germaine Greer and Betty Friedan pointed out that the World
Population Plan up for discussion was written by men and had just one

paragraph specifically about women. Employees of the UNFPA joined protests about India's vasectomy camps. It was a defeat for the population control establishment, a "humbling experience," as Frank Notestein later put it.

After the Indian emergency, this backlash spread. Five days after Gandhi was voted from office, the IPPF—the biggest funder of sterilization in India—agreed that "no sterilization procedure should be performed unless the person concerned has given voluntary unpressurized informed consent." In Sweden, the media dubbed the previous two years "SIDA's Watergate" and forced the agency to stop funding population programs in India. In Iran, Ayatollah Khomeini denounced family planning as an instrument of Western imperialism after he took power from the Shah, an enthusiast.

It might be a stretch to argue that the rise of modern Muslim fundamentalism can be traced to a backlash against Western-inspired population control policies; but those policies certainly didn't help the West's image in the developing world. They even played a role in encouraging the antiabortion movement in the U.S. heartland, a movement that helped bring Ronald Reagan to office in 1980.

· · ·

And yet, somewhere along the way, something worked. The very moment of defeat for the coercive "population control at all costs" wing of the family planning movement was also the moment when it finally became clear that the population boom was past its peak—that the population bomb was being defused. During the 1970s, India's fertility rate started to fall. By the end of the decade, women were having one child fewer than at the beginning. The new Indian fertility rate of 5.4 was the lowest on record. And this was a global phenomenon. Indonesia and Brazil, Thailand and Malaysia, Taiwan and Trinidad, Colombia and Bangladesh, and many more poor countries all started reporting declining fertility for the first time. From the start of the 1970s, the annual growth of the world's population had begun to slacken. From a peak of 2.1 percent, it had shrunk to 1.7 percent by 1984. After years of raising its predictions of future population growth, the UN began reducing them.

CHAPTER 8

Green Revolution

Jitabhai Chowdhury is part of the market economy these days. He used to grow a little sorghum to feed his family on his tiny farm in the Gujarat drylands of India. Now he has a water pump to tap the aquifer beneath his fields and irrigate alfalfa. He feeds the alfalfa to his three buffalo. One morning in 2004 I watched him walk into the village with a churn of their milk to sell to the Amul Dairy. He told me about his two children: a son and a daughter. "I want them to be educated and get jobs. I don't need them on the farm anymore."

Six hundred miles south, near Tiptur in Tamil Nadu, Ratnama told me how she and her husband used to migrate to work on other people's farms because they could not make a living on their own land. Now they are back home growing bumper crops of rice, along with gherkins, bananas, and flowers, on their five acres—crops that they sell in local markets. Ratnama said, beaming, "We used to steal from other farmers just to survive. We went hungry. Now other people steal from us. We have built a new house on the profits and qualify for loans at the local bank."

These are two stories—not special, just typical—among hundreds I have heard across the world, from India and Bangladesh to Indonesia and Kenya, Mexico and the Philippines, in countries with soaring populations that once faced famine and depended on food aid, but now largely feed themselves. They explain in human terms what came to the world's rescue in the last third of the twentieth century—why the dire predictions made by William Vogt at the end of the 1940s never came true; why billions didn't die in famines in the 1980s, as Paul Ehrlich predicted.

Ever since Vogt put down his ornithologist's binoculars and picked up his pen, crop scientists have been galvanized by predictions of world-wide famine. They have seen it as their duty to ensure that every stomach is filled—in a world where more than a billion people are added to the population in little more than a decade. They knew the world could no longer feed itself by ripping up forests, draining marshes, and dragging plows ever further up mountainsides. Their holy grail was a new generation of high-yielding varieties of staple crops like rice, maize, and wheat that could double or triple yields from every field on the planet.

They scoured the world's peasant farms and hedgerows and badlands looking for plants that contain the genetic building blocks to breed these new supercrops. It was almost literally looking for needles in haystacks, but they found them. They found ancient half-forgotten varieties that, if well fed with fertilizer and water, could grow very large heads of grain. And they crossbred them with others that had stalks short enough to hold up those large heads without falling over. That's why the new crops are often called "dwarf varieties."

Who funded this research? Why, our old friends, the heirs to Standard Oil's fortune: the Rockefeller Foundation. In the 1940s, the foundation started paying researchers in Mexico to create better wheat and maize. The now-legendary leader of this enterprise was Norman Borlaug, whose dwarf varieties eventually tripled the country's maize crop and turned Mexico from a wheat importer to a major exporter. In 1967, just as Paul Ehrlich was sitting down to write his population bombshell, the decision was made to take those dwarf varieties out of Mexico and plant them round the world.

Back then, only ten major countries produced more food than they consumed. Most of the rest were kept fed with imports from the United States, Canada, and Australia. The first task was to make more countries self-sufficient in basic foodstuffs. To that end, the first destinations for the Mexican supervarieties were the fields of India and Pakistan, where drought was at that moment turning to famine.

Would the new varieties work in Asia? Ehrlich acknowledged in *The Population Bomb* that "the highest potential for reducing the scale of the coming famines involves the development and distribution of new high-yield varieties of food grains." But he was unconvinced that it would happen. Luckily, his pessimism proved misplaced. By the mid-1970s, wheat and maize yields had doubled in India too. And Borlaug had picked up a Nobel peace prize.

At the same time, scientists tried to repeat the trick with rice. In 1960 the Rockefeller and Ford foundations set up the International Rice Research Institute outside Manila, the capital of the Philippines. They crossbred a big-grain rice variety from Indonesia with a dwarf-stem variety from Taiwan. The resulting IR8, or "miracle rice," was rapidly planted across the paddies of Asia. In the Philippines, the head of the National Rice Sufficiency Program, Rafael Salas, won plaudits for keeping production ahead of a soaring population. The new rice secured reelection in 1969 for President Ferdinand Marcos, who became America's anticommunist bulwark in East Asia for the next two decades. And thanks to the Rockefeller connection, it got Salas a job in New York running the UN Fund for Population Activities. Asian rice bowls overflowed. In the thirty years from 1963, food output outstripped population growth by 36 percent in Asia as a whole, and by 47 percent in East Asia.

Washington was delighted, for—although the story is little told—the United States was using the green revolution to prevent red revolution. "The rice program in particular originated from American anxieties about the possible spread of Chinese communism," says the Canadian environmental historian John McNeill. The U.S. Agency for International Development (USAID) rushed the products of the green revolution around "the frontiers of the communist world, from Turkey to Korea, as a means to blunt the appeal of socialist revolution." The new varieties, it was argued in Washington, would make people richer and happier and head off the much-feared socialist response to rising populations: the redistribution of land from large estates, many owned by U.S. corporations, to peasant farmers.

Seen in this light, the green revolution and population control were both part of a fix to preserve the capitalist status quo. Certainly, green-revolution rice and a crackdown on population growth went together in most countries under U.S. influence. The World Bank, especially once it was headed by the former U.S. defense chief Robert McNamara, invested heavily in rural roads and IUDs, irrigation dams and condoms, fertilizer plants and laparoscopies. Nobody considered the two strategies as alternatives. Borlaug, accepting his Nobel prize in 1970, warned about "the population monster" and said that "the frightening power of human reproduction must also be curbed; otherwise the success of the green revolution will be ephemeral only."

This everyday story of Pax Americana is, however, not the whole truth. Apparently unknown to Western plant researchers at the time, the Chinese were ahead of them. Chairman Mao's plant breeders were much mocked for their devotion to the crazy theories of the Russian scientist Trofim Lysenko. But they too wanted more rice. And in 1959, before the Filipino rice research lab was even open, the Chinese came up with a high-yield variety using the same Taiwanese "dwarfing" gene. The hybrid was growing over millions of acres of southern China in 1965, a year before the release of IR8.

Whatever the politics, and whoever discovered what first, the green revolution kept the world fed through the decades of maximum population growth. China reached a billion people in 1980, twenty years earlier than Vogt predicted, but still managed to feed itself—something Vogt had said "would certainly be impossible." Vogt had said of India that "in all the world, there is probably no region of greater misery, and almost certainly with less hope." It too has doubled its population, yet also doubled its food production and life expectancy.

We should cheer. With one bound, the world was free of its Malthusian, Vogtian, Ehrlichian bonds. Within two decades, three-quarters of the developing world's farmers were using the new varieties. Global food production doubled long before the world's population did. Only Africa, whose local crops never got the green-revolution treatment, saw production fail to keep up with population. The average number of calories supplied to people in poor countries increased by one-quarter, from 2,100 kilocalories a day to 2,650, compared with an estimated need of 2,500.

Yet some do not put out the flags. It seems perverse. But early on in the green revolution, scientists made some choices about the kind of revolution they were going to deliver that, apparently, were the wrong ones. Two decisions stand out. First, they decided to breed crops that would produce very high yields on good soils in response to high inputs of chemical fertilizers and, where necessary, irrigation water. Second, they developed varieties that could be mechanically harvested.

This strategy was, on the face of it, odd. The new crops required heavy capital investment to get the most from them, but needed little manual labor on the farm. Yet poor farmers in developing countries did not have capital, while—thanks to fast population growth—they were awash with labor. The folly of this mismatch was argued at the time. The Swedish economist Gunnar Myrdal said Asian agriculture needed to become more labor-intensive, not less. But the die was cast. As Jack Caldwell puts it, "After the green revolution, you didn't need kids to work the land. For most farms, two or three people working all year can do the job."

Two things happened as a result. First, those made redundant by the new farming took to the road. They filled the burgeoning Asian mega-cities, which, until recently at least, have been kept fed with cheap green-revolution crops. Second, by rating cash over labor, the green revolution discouraged large families. Kids had no place on green-revolution farms. Rather than providing cheap labor, they soaked up cash badly needed to invest in the new type of farming. What looked like a flawed strategy may be seen as genius. Many researchers, including Caldwell, believe the green revolution has been one of the major drivers of falling fertility in Asia, because it literally made children redundant.

Nonetheless, the collateral damage has been huge, especially to the environment. By encouraging farmers to stop planting their local varieties of staple grains, the green revolution has left the huge expanses of high-yield monocultures much more vulnerable to pests and diseases. Right now a virulent new strain of a disease called stem rust is wrecking wheat fields round the world. So far the main response to any new threat has been more pesticides—pesticides that are thought to kill about twenty thousand people a year and pollute the natural environment from the tropics to the Arctic Ocean. But these pesticides produce diminishing returns as the pests evolve to withstand the chemicals.

Meanwhile, the new crops are mostly very thirsty. Their yield for every ton of irrigation water is often lower than that of the traditional crops they replaced. The world is growing twice as much food as it did a generation ago, but taking almost three times as much water to do it. That has required the damming and diversion of hundreds of rivers across the tropics. The result is dried-up rivers, from the Rio Grande in Mexico to the Indus in Pakistan and the Yellow River in China. Water, rather than land, is the new constraint on food production in much of the world today.

Many people I know regard the green revolution as a disaster. They say it has tied billions of the world's peasants to a marketized, globalized, mechanized, energy-guzzling, climate-warming, biodiversity-destroying way of feeding the world. I see their point. And it might have been done differently. But would they prefer billions starving? The green revolution, flawed as it was, kept granaries full, even as population continued to grow at a rate never seen before and unlikely ever to be seen again. And except where people were too poor to buy the food available in the stores, it kept the world fed. I don't discount the fact that there is still malnourishment in some green revolution countries. The most recent figures show 46 percent of Indian children malnourished. But this is a measure of weight, not starvation. It is hugely better than the outright famines of the 1960s.

Whenever famines occur today—mostly in Africa now—the problem is rarely an absolute shortage of food, but rather one of poverty. There is usually food in the granaries, but those who have lost their crops cannot afford to buy it. The poverty is a disgrace, but the production of food is a triumph. The world's food supply system is far from perfect. But to call the green revolution part of the problem seems to me a luxury that would not be endorsed on most farms across the world. Certainly not by Ratnama and Jitabhai.

And along the way, the green revolution gave billions of people access to the cash economy for the first time—making them able to buy TVs and refrigerators and motorcycles and fast food and cheap fashions and mobile phones and much else. If you think that is bad, you have to be willing to give up your own.

· · ·

For a couple of decades, the green revolution took the wind from the sails of demographic doomsters. In their place, the supercrops and the world they created spawned superoptimists, cornucopians like economist Julian Simon of the right-wing Heritage Foundation in Washington. Simon had the enthusiasm of a convert. In the 1960s he was an expert on mail-order marketing and had once sold an improbable 200,000 copies of a book on the topic. After reading Ehrlich's *Population Bomb*, he wrote pamphlets on how marketing could persuade women to have fewer babies.

But this outrider of the population control movement changed his mind about whether more babies were a planetary peril. He figured that if rising populations created resource shortages, that ought to result in technical innovations that fixed the shortages. When the English ran out of timber in the eighteenth century, he said, they didn't go cold. Instead, they began burning coal and started the Industrial Revolution. When whale oil got scarce in America in the mid-nineteenth century, people started drilling for petroleum to light their lamps. When the oil runs out, he reasoned, we'll find something better. Necessity is the mother of invention. Not only that: the greater the necessity, the greater the invention.

Through the 1980s and 1990s, Simon won a huge fan club for his argument that there were no true limits to growth—that Malthus and his modern acolytes, like Ehrlich and Meadows, had it all wrong. That population growth may bring more mouths to feed, but it also brings "more hands to work and brains to think." That, as Nicholas Eberstadt, in-house demographer at the neoconservative American Enterprise Institute in Washington, puts it today, "in the final analysis, the wealth of nations in the modern world is not to be found in mines or forests or deposits of natural resources, but in their people—in human resources. And human beings are rational, calculating actors who seek to improve their own circumstances—not heedless beasts who procreate without thought of the future."

Simon rearranged some ideological furniture. In Malthus's time, free-marketers were pessimists who feared and demonized growing populations of the poor, while social reformers were optimists who saw Malthusian thinking as designed to defend the iniquities of the market. That was still broadly the breakdown in the 1970s. But in the 1980s, free-marketers like Simon were the optimists who wanted free procreation as well, while the social reformers were the pessimists who saw Malthusian limits loom large. That doesn't make either of them right. But it is interesting.

Simon's optimism gained ground as harvests grew and commodity prices sank. Reaganomics seemed to deliver, and Malthus looked less like a sage ahead of his time and more like a fusty old cleric from two centuries before. Simon told me in 1985, "Things have carried on getting better and cheaper for centuries. There is no reason to believe this will stop. The oil crisis [of 1973–80] was a hiccup, not a sea change. Natural resources are not finite, in an economic sense, and population growth is likely to have a long-run beneficial impact on natural resources."

Greens laughed. In 1980 Ehrlich challenged Simon to a high-profile wager. He bet that prices of five metals—chrome, copper, nickel, tin, and tungsten—would rise in the coming decade; Simon bet they would fall. Simon won on all five metals—during a decade in which the world population grew by 800 million and the global economy boomed. He won because plastics, fiber optics, ceramics, and aluminum all stepped in to reduce demand for Ehrlich's chosen metals. Their critical shortage had proved an illusion, as Simon had predicted. In 1990 Ehrlich posted a check for $576.07.

But the argument was not just about metals. On food production, Simon cited an obscure Danish economist called Ester Boserup. Back in the 1960s, after two decades of consultancy work for the UN on farms in India and across Africa, Boserup had argued in an equally obscure book called *The Conditions of Agricultural Growth* that population growth drives technological advances in agriculture that then sustain the bigger population. From harnessing fire and making axes, through the plow, the horse collar, crop rotation, tractors, inorganic fertilizers, and the seeds of the green revolution, farmers had always delivered. Boserup saw no limits to innovation, so no limits to growth. "My conclusion was the opposite of the general opinion at that time, when it was believed that the carrying capacity of the globe was nearly exhausted," she wrote later.

I first stumbled on Simon's book *The Ultimate Resource* in a book-shop in 1984. I was preparing to report on a world population conference in Mexico City. Hard as I tried, I couldn't see the logical flaw in his argument. I did my own research, checking population trends and wealth in the world's poorest countries. There was no discernible link between the two. Africa's ten richest countries had roughly the same population growth as the ten poorest. I ended up writing a magazine feature with the cover line "No Limits to Growth."

Ehrlich says the limits haven't gone away. The crash will come one day. The green revolution just postponed it. The past is not necessarily a guide to the future. And once tipping points are crossed, there is no go-ing back. He compares Simon to a guy who jumps off the Empire State

Building and shouts halfway down that everything is going fine. Simon died in 1998 at the age of sixty-six. Ehrlich, who was born just three months after him in 1932, is, at the time of writing, still going strong. And so is the argument. I will return to it.

CHAPTER 9

One Child

China's population policies of the past half century can be summed up in the final years of one man: Ma Yinchu. The son of the owner of a small rice-liquor distillery, he already had a hugely distinguished career as an economist when, at the age of sixty-nine, he became the first communist president of Beijing University. In July 1957, after being in the job for six years, he was invited to make a presentation to the National People's Congress, the highest state body in China. He called it *Xin renkoulun*, or New Population Theory. The editors of the *People's Daily* considered the report important enough to reproduce it at length. Then the night soil hit the fan.

Carefully phrased and peppered with qualifications, Ma's report was nonetheless a direct assault on one of the most cherished beliefs of the country's leader, Mao Zedong—that more Chinese people would make a better, more powerful China. A couple of years before, Ma had made a pilgrimage to his home village of Pukou, outside Shengzhou, south of Shanghai. He had noticed a huge increase in the number of young children. One of his nephews had eleven children and had got into debt as a result. Back in Beijing, Ma checked the 1953 census returns and discovered that this was a national phenomenon. China had recently seen one of the fastest growths in a country's population ever recorded. Mainland China had 583 million citizens, 100 million more than the previous estimate.

This was a national crisis, Ma told the People's Congress. More did not mean better. China was on course for demographic disaster. It was imperative that Mao abandon his support for large families and persuade his fellow citizens to adopt family planning. Ma didn't want compulsion, but he did want a change of direction.

Perhaps Ma thought his seniority and party loyalty would protect him from the consequences of such a brazen challenge to the Great Helmsman. But soon thousands of posters appeared across the university, denouncing him. "Ma Yinchu claims that people have only mouths; he doesn't see that every mouth has a pair of hands," they said. Gangs of students jostled Ma on campus, and the press called him a capitalist Malthusian. He was forced to resign and went home to write in secret, hoping that the political winds would change. No such luck. Instead he came under siege again in the mid-1960s, as gangs of Mao's Red Guards toured the streets during the Cultural Revolution, attacking intellectuals. Ma, by now turned eighty years old, collected up his papers and burned them all.

China has always been a demographic colossus. For millennia, its fertile soils and rice paddies have sustained a large population, whose energies were controlled by an all-embracing Confucian doctrine of duty and obligation. China has often contained a third of the world's people within its borders. Malthus described China in the early nineteenth century as "the poorest and most miserable of all . . . more populous in proportion to its means of subsistence than any other country in the world." It was a land where periodic famines were "the most powerful of all the positive checks to the Chinese population." A century later in 1926, in his book *China: Land of Famine*, the American geographer Walter Mallory wrote that "the fecundity of these Chinese was without parallel."

In fact, precommunist China wasn't driven by Malthusian forces. A nationwide system of distributing grain meant that famines didn't usually kill many. But the Chinese were expert at managing their fertility, through late marriage, long lactation, periodic abstinence, and all-too-prevalent bouts of infanticide. Decisions on marriage and having children "were exercised largely at the collective rather than the individual level," say James Lee and Wang Feng of the University of Michigan. "Such demographic adjustments allowed them to prosper in spite of China's population size."

By the time of the communist revolution of 1949, the country's population had been stable for about a century. For Mao—who was one of eight children and the father of at least ten, by four wives—this was an affront. He had a soldier's sense that, as one communist slogan put it, "With Many People, Strength is Great." Russian leader Nikita Khrushchev later recalled that while at a conference of communist leaders in Moscow, Mao had said, "If the imperialists unleash war on us, we may lose more than three hundred million people. So what? The years will pass and we'll get to work producing more babies than ever before."

As good as his word, Mao had set about harnessing fertility to the revolutionary cause by first discouraging and then virtually banning contraception. He declared motherhood a patriotic duty. He did this at a time when socialist successes in health were also keeping people alive for longer. It is fashionable today to decry the achievements of Mao's communists, but they had extraordinary success bringing antibiotics and vaccines and sewers to the masses. In three years China immunized almost half a billion people against smallpox and trained three-quarters of a million midwives in sterile techniques. "No longer could sickly children threatened with death be hidden away by poverty-stricken parents," says Jack Caldwell. The results were dramatic. In Mao's first eight years in charge, Chinese life expectancy rose from thirty-five to fifty years.

With deaths down and births up, a staggering number of children were not only being born but also surviving. That was the truth Ma had wanted to get across to his leader. Perhaps only Ma, with more than seven decades of experience, could see the magnitude of what was coming. But far from drawing back, Mao had plans to transform the countryside so it could support the rising population. He reorganized millions of farms into collectives and decided to simultaneously create a rural industrial revolution. He forced tens of millions of peasant farmers to put down their hoes, abandon their rice paddies, and go mining and metal smelting.

This barefoot industrialization, known as the Great Leap Forward, was a disaster. Most of the metal from the farmyard furnaces proved brittle and useless. Drought shriveled the neglected fields. What grain did grow was dispatched to the cities. Soon there was famine. The scale of the disaster did not emerge until after Mao's death in 1976. A team of Chinese demographers organized by Caldwell discovered that for a year or so, almost three times more people died than usual across the country—25 million in all. Meanwhile, 20 million fewer children than normal were registered—some not conceived, some aborted, and some, no doubt, killed at birth. For a few months, more people died in Mao's China than were born. It was probably the biggest famine in history—and almost entirely a product of misguided policies. There was nothing Malthusian about it.

The famine had a sharp global impact. World annual population growth fell from 52 million in 1958 to 41 million in 1959, before picking up to 56 million in 1961. But the effect was short-lived. Within a couple of years, China had a new baby boom. Births were three times deaths during the rest of the 1960s.

China had, as Mao planned, record numbers of hands to work. But they had to be fed. The new green-revolution crops, introduced after the famine, helped. But by the 1970s there was a new pragmatism in Beijing. With Mao's influence fading, a new national population policy emerged. Its slogan was "Later, Longer, Fewer"—meaning marry later, have fewer children, and leave longer gaps between them. Soon birth control clinics appeared, offering women IUDs and sterilization. Fertility began to fall, from six babies per woman through most of the 1960s to less than three by the close of the 1970s. In 1979 the country's new reformist leader, Deng Xiaoping, rehabilitated a frail Ma and made him honorary president of his old university. Ma died in 1982, a month short of his hundredth birthday.

The twenty years during which Ma was banished and Mao's demography ruled saw the fastest population surge in China's history. Numbers soared from 583 million in 1953 to 700 million a decade later, and in 1980 they hit the one billion mark. By the time the country's mandarins realized they had taken the wrong course, they figured the voluntary measures of education and encouragement proposed by Ma would not be enough. After Mao's death in 1976, visitors from the International Planned Parenthood Federation found that provincial family planning committees were setting quotas for IUD insertions and sterilizations. China, a country that seems to find it immensely hard to let people run their own lives, had gone from one extreme to the other. Soon after, Deng introduced the compulsory one-child policy.

Arguably, this policy was simply a reassertion of China's old Confucian controls on childbirth. But in its new state-controlled form, every town, every workplace, every community, every street, and every couple had a quota of babies they were allowed. The careers, and sometimes the lives, of tens of millions of officials, great and small, depended on ensuring those targets were met. Heads were counted, abortions enforced, and even menstrual cycles monitored. It was also a regime in which all the attention was on women. Only 13 percent of Chinese contraception is undertaken by men.

After thirty years, the policy is still largely in place. With time, and in the face of considerable peasant anger, some of its features have been adjusted. While the one-child rule is strictly applied in cities and for employees of many state enterprises, a second child is allowed in many rural regions if the first is a girl and both parents are only children, which of course is increasingly frequent. Minority groups such as the Tibetans are allowed to have three or sometimes more. Only about a third of China's couples today are actually restricted to one child.

In places, and at times, the one-child policy has been vicious, however. In September 1991, a Chinese writer for the *Independent* newspaper in London went with a "2 a.m. snatch squad" to hunt down women who were illegally pregnant. "Their purpose is to make all women who are expecting a second or later child to have abortions and then be sterilized." One village had eleven such women, informed on by locals in return for cash. They were grabbed in their nightclothes, wrapped in quilts, and taken to a local hospital, where the reporter found hundreds of women, many more than six months pregnant, packed in corridors and tents, awaiting their abortions. Some told the reporter they kept getting pregnant because they wanted a boy. Some sympathetic doctors secretly returned to the women any baby boys they induced who were still alive. Next to the clinic was a public toilet. "I went in. There was simply nowhere you could put your feet. It was filled with blood-soaked toilet paper. Behind the toilet stood a line of waste-bins. The aborted babies—some as old as eight months—were put there."

In 2009 reports emerged in the Chinese media that babies born in contravention of quotas were being seized by corrupt local officials and sold to foreign couples. Why don't the Chinese rebel in the way Indians responded to Sanjay Gandhi's excesses in 1977? Sometimes they have. In May 2007 there were riots in the southern province of Guangxi. A family planning official was killed and government buildings torched after local officials, admonished for exceeding quotas, forced women into late abortions and fined hundreds of couples for having too many children.

But with no democracy, local rebellions usually come to nothing. Most people take it for granted that the state can tell them how many children to have. And unlike in India, the state has a reputation for delivering the basics of life, so grudging acceptance of the rules is greater.

An equally interesting question is why the international community backed the one-child policy. Historian Matthew Connelly uncovered a memo written at the IPPF as early as January 1980 discussing the Chinese birth control staff that the federation was training at the time. The memo warned that, once back in China, the staff would be working on "forced family planning, murder of viable foetuses." The memo said that when the scandal emerged, "it is going to be very difficult to defend." The memo was studiously ignored. The IPPF continued blandly to assure questioners that China's one-child policy was "the people's own choice," even though local regulations in at least one province stated that "under no circumstances is the birth of a third child allowed."

Throughout this period, China received funding and support from the UN's Fund for Population Activities. The UN's official policy was that family planning should be "free and responsible." But the free bit seemed to get lost. When demographers at the United States' National Research Council reported mass roundups of women for forced abortions, the fund's deputy in Beijing simply said that the Chinese program was "the only choice for a country with such a large population."

The UNFPA's most high-profile support for China came in September 1983, when it awarded its World Population Prize to the Chinese minister for family planning, Qian Xinzhong. The Moscow-trained former major general in the People's Liberation Army collected his prize in New York, where UN secretary general Javier Perez de Cuellar expressed his "deep appreciation" of how Qian had "marshaled the resources necessary to implement population policies on a massive scale." Qian did have to share the award—but with Indira Gandhi, who a few years before had presided over the other great human rights outrage in the annals of modern population control.

In a bizarre postscript, a British aid minister, Baroness Linda Chalker, revealed years later in answer to a parliamentary question that the Chinese government had "contributed $100,000 to the UNFPA Population Award in 1983," thus, apparently, funding the cash prize to its own minister. It stank.

· · ·

Many countries in Asia took their signal from China and the UN at this time. South Korea adopted a one-child norm. In 1983 Singapore started paying poor Malay women to be sterilized. Bangladesh changed tack too.

When the American feminist academic Betsy Hartmann lived in rural Bangladesh in 1975, family planning officials had refused to supply contraceptives to poor village women. But by 1984 they were forcing IUDs and sterilization on unwilling victims. In Mymensingh district in northern Bangladesh, the army rounded up women in trucks and took them to centers where they were required to sign consent forms before being sterilized. In several provinces, sterilization was a condition for receiving food aid. A quarter of a million operations were performed in the summer of 1984, during the hungry months before the harvest. This was a rate ten times faster than seen before.

Questioned at the time about these outrages, the Bangladesh government blamed overzealous officials. That sounds like a cover-up. It later emerged that a few months before, the World Bank, the UNFPA, and USAID had privately agreed on the need for "drastic" curbs on population in Bangladesh. They wanted the creation of a National Population Control Board with "emergency powers," as well as village visits "from high-ranking government and Army personnel" to get the message across. The World Bank then sent a letter to ministers in Dhaka requiring them to "outline necessary measures so that agreed national population objectives could be met."

In case ministers were left in any doubt about what was required, the UNFPA's local man, Walter Holzhausen, wrote to his bosses that "most donor agencies here greatly admire the Chinese for their achievements, a success story brought about by massive direct and indirect compulsion." The duplicity is breathtaking. All this was weeks before the 1984 UN population conference in Mexico City, where Holzhausen's boss, UNFPA chief Rafael Salas, told journalists that "the UN leaves it to individual countries to determine issues of human rights."

Top demographers sat on their hands, ignoring the UN's behavior. In the run-up to the 1984 conference, Canadian demographer Nathan Keyfitz wrote approvingly in *Scientific American* about China's "most intriguing demographic experiment." All women who had an abortion or sterilization could claim fourteen days' paid holiday, he said, and couples who agreed to have only one child received free medical care and better housing. All carrots and no sticks, it seemed. As in India, only Caldwell stood out. He told a meeting in Mexico City, "I believe in individual rights. I do not believe in coercive programs. I would not have voted the way the committee did on Qian Xinzhong."

But, while most of the world's population planners fell over themselves to defend China's hard-line policies, in Beijing the politicians were having second thoughts. The powerful All-China Women's Federation opposed a policy that, it argued, encouraged women to kill female babies so they could try again for a boy. Shortly after returning from New York with his gold medal and check for $12,500, Qian quietly left office. Rural couples in many areas were thereafter allowed, if they had a daughter the first time, to try again for a son. No thanks to the UN.

· · ·

It is easy to imagine that the draconian powers adopted by communist leaders were accompanied by all-seeing knowledge. Not so. Just as nobody had realized in the 1950s how fast population was rising in China, so later nobody realized how fast fertility rates were falling. Raw census data show fertility plunging to 1.65 children per woman in 1991 and 1.22 in the 2000 census. Zhongwei Zhao, a Chinese demographer now based at the Australian National University, says that the Beijing authorities imagine that their people are keeping babies secret, especially girls. That is why they and the UN still quote a figure of 1.80. But Zhongwei's own cross-checking of census data suggests that fertility really is as low as the census suggests. "It will take time for them to admit they made a mistake," he says.

Something else is becoming clear. Government policies no longer make much difference to how many children many Chinese have. The state still commits despicable acts, but they are no longer necessary, if they ever were, to maintain low fertility. Zhongwei says that in a number of areas where couples have been allowed two children since the mid-1980s, fertility remains lower than in one-child areas.

In large provinces like Guangdong and Liaoning, women on average have *less* than one child. Shanghai's women recorded an average of 0.7 babies in the 2000 census. Xiangyang, a district of Jiamusi city close to the Russian border, has a fertility rate of 0.41, the lowest in the world. The rate in Heping district in Tianjin is 0.43, and in Mawei district in Fuzhou it is 0.46. These figures may be skewed, because migrant workers who appear in the census register babies in their home villages. Even so, says Zhongwei, "in all these cities, there is a one-child policy, but many women don't have children at all."

Perhaps the best proof that the Chinese people don't need government diktats to kick the multibaby habit comes from communities outside mainland China. Throughout Asia, ethnic Chinese women who have never suffered the Beijing yoke are voting with their wombs. Singapore's Chinese women have fewer than 1.1 children each, half as many as Malay women there. Likewise in Taiwan. Most interesting, perhaps, is Hong Kong, the Chinese city-state long ruled by Britain. On the day in 1997 when Prince Charles handed it back to China, its fertility rate was 1.1.

Will the one-child policy be ditched? In the run-up to the 2008 Beijing Olympic Games, an official at the National Population and Family Planning Commission said the government was considering it. "We want to have a transition from control. This has really become a big issue among policy makers." Later, Zhongwei attended a meeting in Shanghai at which "all the demographers said it was time to change the official policy." For policy makers, loosening control went against all their instincts. But in late 2009, the city of Shanghai started actively encouraging couples who are both only children to have a second child, and most China watchers believe this new approach will be extended.

China's population, currently at 1.3 billion, is destined to grow some yet, because the children born during Mao's baby boom are only now completing childbearing. Despite almost three decades of the one-child policy, the number of children reaching adulthood is only now peaking. The resulting momentum will push the Chinese population upward for a decade or so yet. Most guess that it will peak at around 1.4 billion. But after that, the number of women of childbearing age will fall fast, and this vast powerhouse of humanity—something Chairman Mao would probably see as his crowning legacy—will start to shrink. After the boom, the bust.

PART THREE

·

IMPLOSION

Parts of Europe are emptying. Childbirth has dropped to far below the level needed to maintain the population. If Italy gets stuck with current fertility rates, it will lose 86 percent of its native population by the end of the century, and Germany will lose 83 percent. Meanwhile, in parts of eastern Europe death rates are rising, bringing fears for the demographic future of the entire continent.

CHAPTER 10

Small Towns in Germany

Going east, from the bright lights of Munich into Saxony, part of the former East Germany, the train slowed. Outside, the streets looked dirtier and the people older, poorer, and more disheveled. At Plauen—a traditional center of lace making that hosted the first McDonald's in eastern Germany after the Berlin Wall came down—a drunk fell off the train. He dumped his bike in a heap on the platform before haphazardly waving us off. The train rocked on, past boarded-up apartment blocks, sickly yellow trees, and burned-out factories. The only thing that looked new was a sign for a shop that simply said "Bad Service." Even the graffiti looked halfhearted. There were no well-dressed Bavarian city girls left on the train now; old ladies got off in pairs; the carriage was more male as each stop passed.

Even at eight p.m. on a sunny summer's evening, the roads were empty in Chemnitz, an industrial center known for forty years as Karl-Marx-Stadt. The tiny summerhouses on suburban allotments were deserted. I have seen the derelict, rust belt landscapes of former industrial towns before—not least in England, on trains from Sheffield to Doncaster or Birmingham to Wolverhampton. But this world seemed drained of people. In Bavaria, I had asked if anyone ever went to Dresden or beyond. Most shuddered at the idea. I could have been asking about Chernobyl. Of course there were people about, but far fewer than there once were. Out east, the fatherland was becalmed.

I was destined for Hoyerswerda, a town two hours beyond Dresden, close to the Polish border. It has lost half its population in the past twenty years. The young and those with qualifications have left—young women especially. Those that remain have almost given up having babies. The place is becoming an aging ghost town. The main municipal activity is tearing down the bleak apartment blocks of the old socialist utopia. Hoyerswerda is giving street after street "back to nature," though nature shows a reluctance to recolonize the concrete. This isn't the energizing reunification, the "blooming landscape" that Chancellor Helmut Kohl promised after East and West Germany were joined in 1990. Hoyerswerda (known to its citizens as Hoy Woy) seems a town without a purpose, in a corner of Europe without a future.

On the windswept roof of the Lausitz Tower, the town's only landmark, I met Felix Ringel. A young German anthropologist studying at Cambridge in England, he passed up chances taken by his friends to study the rituals of Amazon tribes or Mongolian peasants. As we surveyed the empty plots of fenced scrub below, he explained that the underbelly of his own country seemed weirder and far less studied than these exotic worlds. I rather agreed. And the more I thought about it, the less like the past it looked—and the more like the future.

In its heyday in the 1960s, Hoyerswerda had been a model community in communist East Germany, a brave new world attracting migrants from all over the country. They dug brown coal from huge open-pit mines on the plain around the town. There was good money and two free bottles of brandy a month. Many more worked at power plants that burned the coal and sent electricity as far as Berlin. Hoy Woy pioneered a new design of prefabricated concrete apartment buildings called *plattenbauten*. They were the forerunners of thousands of blocks built across the whole of Europe, west as well as east. The town was also proud to have East Germany's first shopping mall and streets named after Soviet cosmonauts.

But the collapse of communist East Germany changed all that. It was here in 1989, in the towns and cities of Saxony, that the people of the east started moving west to capitalism and freedom. Sealed "freedom trains" carried fleeing Poles and Czechs west on the Dresden–Leipzig railway. Hundreds of thousands demonstrated in Leipzig, Dresden, and Chemnitz against the Honecker regime. Then, after the Berlin Wall came down and communism fell, eastern Germans too flooded west.

At the head of the queue were the young, and especially young, women. Under communism, East German women worked more than the more conservative western hausfrau. They also were often better educated. When their jobs disappeared in the early 1990s, hundreds of thousands of them, encouraged by their mothers, took their school diplomas and CVs and headed for the west to cities like Heidelberg. The boys, seeing their fathers out of work, often gave up. Today there are twice as many eastern German male school dropouts as females. In adulthood, they form a rump of ill-educated, alienated, often unemployable men, most of them unattractive mates—a further factor in the departure of young women.

Reiner Klingholz, director of the Berlin Institute for Population and Development, calls it a "male emergency." He has a bleak map of the sex ratio of eighteen- to twenty-nine-year-olds across Germany. At the border between west and east, the color changes from the blues and pinks of a female majority in the west to the dark reds and browns of the east, where men are in the majority. In Heidelberg in the west, there are today 122 young women for every hundred young men. Outside Berlin, most eastern towns have fewer than eighty women left for every hundred men.

But this is not just an emergency for men, for the exodus from the east is only part of the story. That may end now that the economic bubble in the west has burst. What looks less likely to end is the collapse of childbearing across eastern Germany. The former people's republic is staring into a demographic abyss, because its citizens don't want babies anymore.

After the Berlin Wall came down, millions of eastern Germans who stayed behind decided against producing another generation. Their fertility decreased to less than half its former level. In 1988, 216,000 babies were born in East Germany; in 1994 just 88,000 were born in the region. In the south, in Brandenburg and Saxony, baby making fell by two-thirds. No demographers predicted it. "There has been nothing comparable in world peacetime history," says a French demographer, Jean-Claude Chesnais. The fertility rate worked out at 0.8 children per woman. Since then, it has struggled up to around 1.2, only just over half the rate needed to maintain the population.

Because of outward migration and falling fertility among those who have stayed behind, virtually the whole of the former East Germany is losing population. About a million homes have been abandoned. The government is demolishing them as fast as it can. More than 400,000 were scheduled to go by the end of 2009. Left behind are "perforated cities," with huge random chunks of wasteland. Europe hasn't seen cityscapes like this since the bombing of the Second World War.

Magdeburg and Chemnitz, both large industrial cities, have lost a fifth of their population. The bars and cinemas and offices have all shut, along with the factories. "Some towns are losing their purpose," says Katrin Grossmann, a Leipzig geographer. Halle has lost 23 percent of its population and Gorlitz 24 percent. Gera and Suhl in Thuringia; Cottbus near the Polish border in Brandenburg; Dessau in Saxony-Anhalt; and Neubrandenburg near the Baltic coast—all have lost more than 15 percent. Erfurt, the capital of Thuringia, has half as many children as in 1990.

But no town has emptied as much as Hoyerswerda. In the 1980s it had a population of 75,000 and the highest birth rate in East Germany. But today, with the mines and power stations shut, the town's population has been reduced by half. Hoy Woy has gone from being Germany's fastest-growing town to its fastest-shrinking. Once with the youngest population in East Germany, it is now has among the oldest. The town authorities say there are seven thousand fewer families. Surplus apartments are being demolished. Those citizens who remain huddle together in clots of urbanity. A quarter of them are unemployed.

Most of the schools have closed. In twenty years the number of children under fifteen has fallen by 80 percent. The biggest age groups today are in their sixties and seventies, and the town's old birth clinic is now an old people's home. Its population pyramid is upturned—more like a mushroom cloud.

The pockmarked plain around the town, which once hummed with industrial activity, is now deserted, apart from a clutch of wind turbines that beat mockingly on the horizon. The mayor of Hoyerswerda, Stefan Skora, hopes to halt the population slide at thirty thousand. The town's Web site talks brightly of a "town in transition." The old coal mines will be flooded and turned into "the biggest artificial lake district in Europe," a holiday camp for water sports. But who will come?

There are still nice people here who try to preserve the old sense of community. In a school in a partly demolished suburb known even to its inhabitants simply as Area Nine, I met Nancy, a tattooed and quiet-spoken social worker. Forty years ago, Nancy's parents were among the newcomers. Her mother was a midwife and her father a train driver. "We came from Magdeburg. There were modern flats and services here then. It was a prestige development, a modern socialist city." But now? It's not just the infrastructure, it's the people, she said.

"Teachers here have lost their moral influence within the community. They are just educators now. But the children need more. Once, when you asked the kids what they wanted to do when they grew up, they had ambitions to drive buses or work in the power station. Now they joke, 'I'll go to Netto.'" Netto is the local discount store where the alcoholics hang out and the neo-Nazi skinheads prowl. "Parents find it very difficult to encourage their children to think ahead when they have no jobs or prospects themselves." Those who can, go. "My friends have all left. I'd like to stay, but I have a three-year-old daughter and the schools are no good anymore. I'll probably go too."

Further out, in Area Ten, I came across Marco. He's twenty-seven. "Only criminals live in this neighborhood now," he said. It was so rough, even the bar had recently closed. Neo-Nazis had taken over the Christian youth club. Marco, the child of an alcoholic mother and a violent father, spent five years in the town orphanage and was doing odd jobs to pay off debts. As we shared a Coke, his face was scared and exuberant, fragile and intense, delicate and feral. "I don't know who I am yet," he said. "I've never experienced a family. I'd love to have my own. But this place is empty for me. I get so angry, I thought of slitting a woman once. I'd like to go to America when I am out of my debt; that's my dream." The dream of a doomed man in a dying town.

Strangely enough, in this adversity, some kids show spirit. Felix took me to meet artists, photographers, rock musicians, and graffiti artists at the town's arts center, known as the Kufa. Local heroes include Hoy Woy's answer to Bruce Springsteen. Gerhard Gundermann was an excavator operator at one of the mines who became a melancholic singer-songwriter. He died at age forty-three. Ten years later, the Hoy Woy counterculture still reveres him. But the city fathers hate him. He was in the Stasi, they say. But then a third of the country—and no doubt a third of Hoy Woy's citizens—were signed on to the secret police's fabled network for spying on its own citizens. It was a passport to survival—especially, perhaps, among cultural dissidents.

Anyhow, the kids don't care. At the Kufa they have made a street sign in Gundermann's name. They want to put it up in place of one of the cosmonauts' signs. But even in a post-Soviet world, the city won't let them. "For us, nothing has changed," said Uwe Proksch, the bearded, laid-back boss of the Kufa. "We are still the outsiders, still the subversives. The city authorities think we are weird people doing odd things."

Marco stays. He probably always will. But with no jobs or prospects, the smart kids depart as soon as they can. "There is no reason to stay," said high school student photographer Florian, as we looked at his moody photos of the town at night. His friend Benny, a graphic artist, was bound for art college in Dresden. It is where a lot of the Hoy Woy boys end up. Even the prostitutes went there, says Felix. By order. "Brothels are allowed only in towns with more than fifty thousand people. We are below that now." Sex, like much else, is now for sale only in Dresden, two hours away by train.

Of course, rust belt cities have gone into decline before, hollowed out by industrial decay and the flight to the suburbs. But normally, suburbanites commute to the city for office work and entertainment. And their children move into city-center apartment blocks when they grow up. The cities transform and renew themselves. But in eastern German towns today, there is simply nobody left to rebuild. The total population of the former East Germany has fallen from 16.7 million to 13.6 million since reunification.

Communities are falling apart. Racism is endemic in the clutch of small towns between the big Saxony cities of Dresden, Leipzig, and Chemnitz. Wurzen is home to a mail-order company selling clothes and music extolling the far right. One summer night in Mugeln, a row on a dance floor escalated into a crosstown manhunt. Neo-Nazis pursued Indian youths to a pizzeria, which they ransacked, leaving fourteen people injured.

Most famously, there is Hoyerswerda. Back in 1991, a group of more than two hundred refugees from Mozambique partied long into the night in their hostel in the town. Sleepless locals got angry. Soon there was a fierce standoff. The police didn't know what to do. In the chaos of reunification, there were no politicians to tell them. Things escalated. The hostel was besieged for almost a week before the military came to the rescue. Across Germany, it is the one thing most people know about Hoyerswerda. It reinforces the image of parts of Saxony as a feral wasteland.

Feral enough for wolves. Crossing the borders from Poland and the Czech Republic, the new inhabitants have been arriving for several years now. Slinking into the depopulated towns and villages of Saxony, wolves are finding empty spaces where once there were apartment blocks and mines. They are staying. A few miles down the road from Hoyerswerda, near the tiny town of Spreewitz, wolf enthusiast Ilka Reinhardt can't believe his luck. "We have more wolves than we have had in two hundred years," he told *Der Spiegel* in 2007.

He spends nights driving through Saxony trying to follow the signals given off by radio collars he has fitted to a handful of the beasts. The badlands of the former East Germany are indeed going "back to nature." And Europeans should be worried, for some fear that eastern Germany is, as it was back in the sixties, a trailblazer for the demographic future of the continent.

CHAPTER 11

Winter in Europe

They call it Europe's demographic black hole. Doomsday for a continent. And it was entirely unpredicted. Sixty years ago, William Vogt forecast that Europe could die out because it would not be able to feed itself. But now it looks like a birth dearth is about to plunge the continent into a tailspin of ever-declining numbers. Europe's population is right now peaking after more than six centuries of continuous growth. Yet with each generation now reproducing only half their number, what is happening in Hoyerswerda and eastern Germany looks like just the start of a continent-wide collapse in numbers. Some predict wipeout by 2100.

Half a century ago, Europe was basking in a postwar baby boom. Fertility rates were 2.8 in Britain, 2.9 in France, and 3.2 in the Netherlands. Then levels sank back. Demographers assumed fertility would settle down at around the level needed to maintain the population—a bit over two babies per woman. That is what Frank Notestein's theory of demographic transition predicted.

The trouble was that nobody told Europe's women. In the real world, the swinging sixties saw a great deal of sex and not a lot of procreation. At the start of the decade, before the Beatles, European women on average had their first child at age twenty-four. But that number was rising steadily, and by the 1990s, the first booties and shawls showed up at twenty-nine. The maternity wards and primary schools were emptying. Women had extracted the best part of a decade of freedom from diapers and breast-feeding and kindergarten and the school run. Instead, they were going to universities in ever-greater numbers, and most of them were then getting jobs.

Demographers held conferences to discuss the new phenomenon. At first they called it the "tempo effect." Don't worry, they told the planners. Childbirth was now something women do in their thirties rather than their twenties. There might be a delay in starting families, but it's only that. The postwar template of Mum, Dad, and the two or three kids—the nuclear family reprised in a million TV commercials and sitcoms—was still the norm. Wrong. Whether women had planned it all along or just never got round to turning sex into babies, more and more of them stopped at one child, and many were not giving birth at all. "Waiting for tempo is beginning to look like waiting for Godot," says Peter McDonald of the Australian National University in Canberra, who watched the scenario play out in Sydney, as it did in Stockholm and Seattle and the Sorbonne.

By the mid-1980s the alarm bells were ringing. "Europe is entering a demographic winter," declared a French demographer, Gerard-François Dumont. Robert Lesthaege at the Free University in Brussels blamed "post-materialistic values, in which self-development becomes the primary aim." Michel Debre, a former French prime minister, went further: "We have destroyed marriage and the family. We have erected greed and self-indulgence as our idols."

This was a moral panic with geopolitical force. For hundreds of years, Europe has exported its ideas and products, and especially its people. Force of numbers mattered as it colonized the planet. But a resolution at the European Parliament in 1984, calling for action, warned that Europe's share of the world's population was set to drop by half between 1950 and 2000 and was likely to be halved again as soon as 2025. The "trend will have a decisive effect on the significance of the role which Europe will play in the world in future decades." Europe's population in 1950 was double that of Africa, over which it then largely ruled. By 2050, according to UN predictions, Africa would have a population three times greater than Europe. The thought of Africa ruling Europe was too horrible to contemplate.

Some believe this is all far-fetched—that recovery could be just round the corner. Britain's population is still rising. But there are long-term trends that cannot be escaped. The twentieth century began with western Europe producing ten million babies a year; by the end it couldn't manage six million. That was two million fewer than needed to maintain the population in the long term. The baby famine is now heading into a second generation. It is no longer a blip. Demographically, Europe is living on borrowed time.

German, Austrian, Russian, Swiss, Spanish, and Greek women manage just 1.4 children each; Italians, 1.3; and Czechs, Poles, Bosnians, Ukrainians, and Belarusians, 1.2. Thirty years ago, twenty-three European countries had fertility above replacement levels; now none do. Only France, Iceland, Albania, Britain, and Ireland are anywhere near. In those thirty years, seven European countries have halved their fertility. In the mid-1970s, the continent's least fecund women were Germany's at 1.5. But by 2008, twenty-five countries had fertility rates below that figure.

These are probably the lowest rates of fertility in nonsuicidal populations in history. The only time like it was after the First World War, when a generation of young men died on the battlefield. But today there is no shortage of lusty men—only of broody women. Demographers expect a small bounce as childless women decide to have late babies before the menopause. But it will be small. In almost every European country, most women are having at least one fewer child than their mothers and two fewer than their grandmothers.

Now the economic downturn threatens to depress fertility further. "The financial crisis will put a major burden on families, and they will reduce costs and also the number of children," said German social anthropologist Wassilios Fthenakis in early 2009. "There is a good bit of evidence that hard economic times cause people to delay having babies or not have one altogether," agrees Carl Haub, senior demographer at the Population Reference Bureau in the United States. Whatever the short-term prognosis, however, it is the long term that matters. And the new baby bust has profound implications, not just for Europe but perhaps for *Homo sapiens* as a whole.

. . .

Let's start with Isabella. She lives in Rome. She is in her thirties. She has a nice apartment, a nice job, and a nice boyfriend. He still lives with his mother, as most Italian males do until their late twenties. So Isabella can see him when she likes and keep away when she doesn't. Isabella is happy with her childless life. With a child, she would probably have to give up the apartment. Her employers, who don't have any child-care provision, would swiftly find a younger replacement. And so might her boyfriend. Even if he did agree to move in and help with the baby, he would expect the same housekeeping service he gets from his mother and would be hopeless at changing diapers. He's better on tap than on top.

Isabella's sexual and working lives could hardly be more different from her mother's. Today's young Italian women spend an average of nine years after their first sexual intercourse before marrying, compared to less than two years a generation ago. And thanks to the spread of the pill and condom, they are much less likely to get pregnant. Italy is the home of the Catholic Church, with its fundamentalist opposition to artificial birth control. But it is a church that has for the past thirty years presided over the emergence of one of the lowest fertility rates in the world.

And it is not just an urban phenomenon. For while Isabella is the model of a modern urban working girl, too busy for babies, what are we to make of her older sister Clara? Back home on the Mediterranean island of Sardinia, Clara is married but not working. She is one of the 40 percent of women in southern Italy who have never had paid employment. But like millions of other young married women, she too is childless. She has the time and the husband, but not the inclination. She says they are too poor to start a family—something her mother would never have said.

For the first time in Italy's history, the churchgoing, confession-observing south has lower fertility than the north, says Francesco Billari, professor of demography at the University of Bocconi in Milan. Nowhere is this more obvious than in Sardinia, where in 1960 women had more babies that anywhere else in Italy, an average of 3.5 each. But now they have the fewest, just 1.1. As a result, houses and farms lie abandoned, schools are empty, and whole villages are dying.

Italy has had below-replacement fertility for thirty years now, and there's no sign of recovery. With few babies, it has gone from having one of the youngest populations in Europe to having one of the oldest in the world—with the highest pension bill in the European Union. At the time of writing, its prime minister is Silvio Berlusconi, who at seventy-one is Western Europe's last national leader born before the Second World War. But Italy's people crisis is being replicated across much of Europe. I sat in on a meeting of Europe's top demographers in Barcelona to discuss what Billari was the first to dub Europe's "lowest-low" fertility. They said that, probably for the first time in history, not having children is both a widespread and socially acceptable lifestyle option in much of Europe.

Some at the meeting blamed lowest-low fertility on a "Madonna syndrome"—not the Madonna mother venerated by the Catholic Church, of course, but the modern "material girl." Others criticized the "Bridget Jones generation" of working women with complicated love lives who dither about choosing men. Still others said it was the men themselves. Why would modern women want to have babies with undereducated, conservative, and socially maladroit "loser men"? Men who were cowed by the liberation of their womenfolk, joined neo-Nazi parties in eastern Germany, sponged off their parents in Italy, and flipped burgers in Britain?

In Italy, there may be another factor: the mother-in-law. Italian families are close. Suffocatingly close, sometimes, says the country's leading demographer, Massimo Livi-Bacci. People in their twenties, especially men, remain in the bosom of their families long after their contemporaries in northern Europe have flown the nest. And as families become smaller, the ties and the suffocation intensify. "In Italy, people love children too much. The paradox is that too much family leads to too few children," says Gianpiero Dalla Zuana of the University of Messina.

"The behavior of young people is alarming," agrees Alessandra De Rose of the Sapienza University in Rome. "There is a delay in all life-cycle stages: end of education, entry into the labor market, exit from the parental home, entry into union, and managing an independent household." It is almost as if today's young adults can find no place in an aging society. Clara fits this category. She feels poor, excluded, and worthless, without the confidence to bring a baby into the world.

Behind the dysfunctional family, said the Barcelona meeting, lie the dysfunctional roles of the state and church. They both promote an old patriarchal ideal of large families in which the wife stays at home. The state denies any responsibility for child care or helping mothers into work. The church, despite losing its influence in the bedroom, retains power over the political climate and public services. This, the demographers said, turns out to be a lethal combination for baby making. Where women are grabbing their new rights but men are not taking their new responsibilities, the result is ultralow fertility.

· · ·

But before giving up on Europe, let's meet Astrid in Stockholm. She has never married, but has a cohabiting partner and two children. She got a year's maternity leave when each of her children was born. She works a flexible thirty-hour week and can put her children in a nursery at the office when she needs to. Her partner, Sven, is adept at changing diapers and takes turns with the four a.m. feeding. Not surprisingly, Astrid and her girlfriends feel more able to have a family than Isabella's crowd, and richer and more assertive than stay-at-home Clara. It shows in the national data. Sweden's fertility rate is 1.7 children per woman.

The Swedish lesson is that where employers, the state, and men are more flexible, national fertility rates are higher. Not at replacement levels, but not set to demographic meltdown either. Those countries are mostly in northwest Europe: in the British Isles, Scandinavia, and France. France, with its thirty-five-hour working week, lavish child-care provision, and traditions of middle-aged female chic, has done best. The Socialist candidate in the 2007 presidential election, Segolene Royal, was a mother of four.

A lot of this comes down to power, says Scandinavia's top demographer, Gosta Esping-Andersen. These days, most couples have a "bargaining process in order to reconcile employment and child care." Women who work, especially those with good jobs, can drive a better bargain. They also have the pick of the available men—choosing those who will change a diaper as well as be good in bed. In Scandinavia, 85 percent of the best-educated women have children, compared with only 60 percent in more conservative and patriarchal Germany. For educated Scandinavian women, having babies is not a duty but a pleasure. A luxury, even.

· · ·

Lowest-low fertility could be a passing phase for Europe. Women in Italy, and other countries with lowest-low fertility, may eventually increase their bargaining power with the state, employers, and men.

Then they might have more babies, like their working sisters in northern Europe. Roll over, Berlusconi. But where the old ways persist, there is a growing risk that societies simply get out of the habit of having children around. Once, the Mediterranean countries were a byword for child-friendliness. In cafés and town squares, children were everywhere. Life revolved round *bambinos* and *bambinas*. Not any longer. And in Germany, where fertility has been low for more than a generation, demographers report a large decline in the desired family size. "Today, 48 percent of German men under forty agree that you can have a happy life without children. When their fathers were asked the same question at the same age, only 15 percent agreed," says Europe's top demographer, Wolfgang Lutz of the Vienna Institute of Demography. Thirty percent of German women today say they don't intend to have children at all.

Combined with this increasing reluctance to have children is a fall in the number of young adults of an age to have children. Once a country has very low fertility for a generation, it begins to run out of young women able to gestate future generations. Germany is there already. It has only half as many children under ten as adults in their forties. Just to maintain the current number of births, those children will have to have twice as many children as their parents. Lutz argues that Germany may now be beyond a point of no return, especially in the east of the country. In another generation, countries like Germany and Italy will have only three hundred women of childbearing age where once there were a thousand. It will hardly matter how many babies these women have—three, four, five—the population still won't recover.

McDonald calculates that if Italy gets stuck with recent fertility levels and fails to top up with foreign migrants, it will lose 86 percent of its population by the end of the century, falling to eight million compared with today's 56 million. Spain will lose 85 percent, Germany 83 percent, and Greece 74 percent.

Jesse Ausubel of the Rockefeller University in New York fears "the twilight of the West" as Europe's population diminishes and ages. "One can imagine a shrinking Europe, whose residences fill with immigrants from North Africa, who spread their culture hostile to science. Civilizations have simply melted away because of poor reproductive rates of the dominant class. . . . The question may be whether underneath the personal decision to procreate lies a subliminal social mood influencing the process. The subliminal mood of Europe and its retinue could now be for a blackout after one thousand years on stage."

Far-fetched? Maybe. But David Reher, a population historian at Complutense University in Madrid, told *Science* in 2006 that "as population and tax revenues decline in Europe, urban areas could well be filled with empty buildings and crumbling infrastructure . . . surrounded by large areas which look more like what we might see in some science-fiction movies." David, come and see Hoyerswerda. The future is already here. Complete with wolves.

CHAPTER 12

Russian Roulette

Boris drunk at a state banquet; Boris slurring his words on TV; Boris stumbling at a parade; Boris leaping onstage to dance out of time and sing out of tune. It is all on YouTube now. But back in the 1990s, watching their president's alcoholism on display was a grim reality for Russians as the country staggered from communism to capitalism. But at least Boris Yeltsin had staying power. He lived to be seventy-six. For most of his countrymen, that is a distant prospect.

Russia is Europe's sickest country. Its people may be freer today than under communism, but they are living less long than they used to. Especially the men. High death rates, combined with low birth rates, are cutting the population of the world's seventh largest country by more than half a million a year. Vladimir Putin, in his presidential address to the Russian federal assembly in 2006, called Russia's demographic crisis "the most acute problem facing our country today."

Russia has been in demographic turmoil for generations. The aftermath of the First World War, the 1917 revolution, and an influenza outbreak the following year left it with six million fewer people in 1923 than in 1914. At least another ten million died in the chaos of famine and Josef Stalin's collectivization of farms in 1932–33—an outcome so catastrophic that Stalin liquidated the statisticians who uncovered it in the 1937 census.

No country except Belarus suffered more than Russia in the Second World War. At least twenty million Soviet soldiers, the majority of them Russian, were killed in the defeat of Hitler's Germany. By the time anyone got round to counting who was left, there were only sixty middle-aged men for every hundred middle-aged women. In the years afterward, the population recovered and grew. For a while, the proportion of men recovered too. But since the late 1980s, male mortality has been rising again, and their proportion in the population has resumed its freefall.

In 2008, Russian women could expect to live to seventy-three. That is poor by the standards of many countries—thirteen years less than Japan, for instance. It is roughly the same life span as women in Thailand. But men? The average Russian man today can expect to live only fifty-nine years. That "three score minus one" is nine years less than in the mid-1980s. The death rate among Russian men in their forties today is double what it was in the 1960s. No other industrialized country has ever experienced such a reverse. And the fourteen-year gap in expected life span between men and women may also be a peacetime world record. On a global ranking of male life expectancy, Russia comes 164th, sandwiched between Cambodia and Ghana. Russian men can expect lives four years shorter than Bangladeshi men, seven years shorter than Indian men, and nine years shorter than Brazilian or Indonesian men.

This is worse than a war. Russian men are poor eaters, frequently violent to themselves as well as others, often sick, and all too often, dead. Why?

Academics do like to beat about the bush. It may be true, as sociologist William Pridemore of Indiana University puts it, that "marketization led to a failure of Soviet state paternalism—including the state's social safety networks and guarantees of medical care, housing, and food—that had disastrous effects for the population." Timothy Heleniak could be spot on when he says, in a study for the Population Reference Bureau, "It appears males in Russia are far more susceptible to economic and social dislocations than are females." But the rot set in before marketization. Before the market revolution came the vodka revolution. It happened in 1988.

During the mid-1980s, the reformist Soviet leader Mikhail Gorbachev had crusaded against alcohol. He shut down state distilleries, banned alcohol sales before two p.m., and raised the price of vodka. The reform, part of his perestroika program, lifted male life expectancy by three years. Then in 1988 the campaign was halted. Illegal distilleries were rapidly replacing state enterprises, and the state was losing tax revenues. As state restrictions eased, the cost of a bottle of vodka went from twice as much as a kilo of sausages to half as much. Russia men responded as might be expected. The binge began. They drank themselves to death in very large numbers. In the next six years, annual male deaths increased by half a million. The life expectancy of the average Russian male fell by seven years.

As millions of men replaced food with alcohol, the World Health Organization estimated that a third of all recorded deaths in Russia were from alcohol poisoning, alcohol-related diseases such as heart disease, or poor nutrition. Accidents and suicides also soared. Pridemore found a tight link in both time and place between binge drinking and rates of homicide. It was a catastrophe. More men appeared to be dying of the effects of drink in the early 1990s than died in the battles of the First World War. In 1992 Russia tipped from being a country with a growing population to one with a declining population. In the midst of all this, Gorbachev was deposed and replaced by Yeltsin. The abstainer was replaced by the alcoholic.

Much changed in Russia apart from the consumption of alcohol, of course. By and large, Russian women have adjusted to the new political and economic order in ways that Western women would recognize. They are having fewer children. They are having them later. They are using modern contraception rather than abortion. They are often giving up on marriage and throwing out abusive partners. Men have coped far less well, and vodka has eased the pain.

I know some of this from personal experience. In 1993 I spent three weeks on a field trip to Siberia with a dozen or so Russian scientists. They were in a bad way. Some hadn't been paid for months. Some hadn't seen their wives for years. There seemed to be very little in the shops apart from ludicrously cheap vodka. The drinking of these learned, erudite men was indescribable. Bottle after bottle. One night, as we camped on a riverbank close to the Arctic Circle, I tried to keep up. It left me in bed for two days and with a hangover that lasted a week. I then flew home. The scientists stayed. Most of them are now dead. Not to put too fine a point on it, millions of Russian men of all backgrounds and qualifications are drinking themselves to an early grave. It is the vodka, stupid.

Under the demographic onslaught of vodka and falling fertility, parts of Russia are emptying. In 2008 *Guardian* journalist Luke Harding visited the village of Slyozi in western Russia. It is just twenty-five miles from Latvia, from the new border with the European Union. He found that the village had shrunk from one hundred people half a century ago to just four—all of them women. The last man had died the previous year. "There are at least 34,000 villages inhabited by 10 people or fewer," he reported. In the next village, Velye, there were a few men left, but the oldest woman, seventy-nine-year-old Zinaida, told him, "Drink is a huge problem round here. The men drink their pensions as soon as they get them. They drink anything—moonshine and even window-cleaning fluids."

In remote areas, this demographic decline is not just a result of death and falling fertility but also of migration. Huge swaths of the Arctic and Siberia are being abandoned as jobs disappear and the infrastructure packs up. You don't want to be stuck in an Arctic city if the district heating system springs a leak. Russia's three great Arctic cities—military Murmansk, metal-smelting Norilsk, and coal-mining Vorkuta—have lost a third of their population since 1990.

Russians, always a nationalistic people, are becoming unnerved by the demographic decline. In the far east around Vladivostok, which has lost a fifth of its native population since 1990, whole districts are being taken over by Chinese migrants. Most are temporary workers for Chinese companies that have bought logging rights in the taiga. But some wonder whether Russia can hold on here, seven time zones east of Moscow. There are 120 million Chinese in provinces that border the Russian far east, which is now home to just seven million Russians.

"Russia for Russians" is a frequent cry. But that is hardly a tenable policy when Russians are a dying breed. One solution may be to import new Russians. "Active acceptance of immigrants offers the only way to slow down or stop the shrinkage of Russia's population," says Anatoly Vishnevsky, director of the Center for Human Demography at the Russian Academy of Sciences.

Where is this headed? Paul Demeny of the Population Council in New York compares the demographic trends of Russia and Yemen. Get this: in 1950 Yemen had fewer than five million inhabitants, compared with Russia's then 103 million. For the Russian bear, tiny Yemen was all but invisible. But Russia is on the slide, and Yemen has the world's sixth highest fertility. On current trends, with rising life expectancy and continued high fertility rates, Yemen's population will reach 103 million by 2050, while Russia's will have slumped back to the same figure. Yemen bigger than Russia? I wouldn't bet on that actually happening, but the prospect must bring Russians out in a sweat from St. Petersburg to Vladivostok.

. . .

Across eastern Europe declining fertility is almost universal, and there is a deep cynicism about state calls for baby making, born of past experiences under communism. Romanians will take a long time to recover from the grotesque population policies of their old communist dictator, Nicolae Ceausescu. He fought the growing reluctance of his citizens to have babies by banning abortion, severely restricting contraception, and declaring "the fetus is the socialist property of the entire society. Giving birth is a patriotic duty, which is decisive for the fate of the country." And when that jingoistic call failed, he ran "birth squads" that visited women in factories and offices, conducted pregnancy tests, and tracked the progress of any fetuses they found.

Outside Ceausescu's madhouse, socialism offered more carrots and fewer sticks to encourage babies. Eastern Europe maintained a welfare state that citizens of the new capitalist east can only dream about today. In Poland there was maternity leave of up to two years. Child care was normal in state factories, and the families of working women could use the subsidized factory canteens. There was little coercive pressure to have children. Abortion was legal (and widespread). Contraceptives, though sometimes of dubious quality, were available and subsidized.

With the fall of the Berlin Wall, state-funded support services for families collapsed. The story of Hoyerswerda—of plunging fertility rates, unemployment, impoverishment, and out-migration—has been played out from Gdansk to Odessa. Poland, in its effort to cut the size of state government, has cut maternity leave to four months and closed workplace nurseries. Employers today reward pregnancy with dismissal, and unemployment among women has risen to twice the rate for men. The resurgent influence of the Catholic Church in Poland has also seen abortion banned, though clandestine abortions proliferate. Newspapers run small ads listing cell phone numbers for euphemistic services such as "recovery of menstruation"—for all the world as if this were Victorian Britain.

In a university suburb of Poznań in central Poland, I discussed this with Marek Nowak of the Polish Sociological Association. The Polish situation is very like that in southern Europe, he said. While public policy is dominated by the Catholic Church, personal life is still largely secular. Vatican-inspired attempts to revive a traditional patriarchal family structure amid the ashes of communism have failed. Since the state and employers alike are reluctant to help women work and have children, most young women are giving up on having children. When Nowak asked Poles their main goal in life, almost three-quarters mentioned work, money, or shopping; only 16 percent referred to home life and only 13 percent mentioned having children. "People are much more interested in self-fulfillment than family now," says Nowak, himself a churchgoing Catholic with a working wife and two children.

Poland, like southern Italy, was until recently a peasant society with large families, he says. "Everyone on this campus has a grandparent who was a rural peasant. We only industrialized and urbanized after the Second World War." Half a century ago, most women had four children. Before communism collapsed they typically had two, and now fertility is 1.3, with urban areas reporting just 1.1. The government has called for women to do their duty and have more babies. But one woman in Gdansk responded with heavy sarcasm: "I am happy we are having a demographic crisis. It's what the government deserves."

"Lowest-low fertility is unlikely to be a short-term phenomenon in eastern Europe," says Italian demographer Francesco Billari. "It is likely to persist at least for several decades." Whatever happens now, the next generation of baby makers will be substantially smaller than the last. Falling populations are now probably hardwired in much of eastern Europe.

PART FOUR

·

THE REPRODUCTIVE REVOLUTION

Low fertility has spread from Europe to become a global phenomenon. Rich and poor, urban or rural, Islamic or Catholic, capitalist or socialist, educated or illiterate, women everywhere are giving up on large families. Because they can. Because for the first time in human history, most babies make it to adulthood. Only in a few patriarchal backwaters is the trend being held back. The repercussions of this reproductive revolution are profound—and sometimes unexpected.

CHAPTER 13

Sisters

Akhi is a sweatshop worker, employed in one of the thousands of garment factories in Dhaka, the capital of Bangladesh. I met her one evening in her one-room corrugated iron shack, down an alleyway in a slum called Mohakhali, famous for its flyover, the fast-growing megacity's first highway overpass. Sitting beside her on a bed beneath a whirring fan were fellow workers Aisha and Miriam, who were sisters-in-law. The three shared the room with two other women who were still at work. Two beds between the five of them.

The young women in their cheap saris told me about their families. Akhi had seven sisters and brothers back in her home village of Kumila, just outside Dhaka. The family rice farm could not support them all and did not need their labor. So she had come to Dhaka to help make the world's clothes at rock-bottom prices. She received five pence an hour for sewing collars onto shirts for ten hours a day—overtime compulsory. Aisha had four siblings, and Miriam had six. All three sent money home from their paltry wages. Their families depended on them.

Would they have children themselves? Two of them already did. Akhi had a son and would like one more child. "I have kept my son in the village for now. I may go back. But it might not be possible, so I may stay and bring him here." Aisha had two sons and a daughter, all looked after by her mother. "But that is enough. I don't want any more children." Miriam, still childless, "would like a child one day, but only one."

In this one room, I realized I had witnessed a demographic transformation. These three young women, from families totaling twenty children, planned just six children between them. The remarkable truth is that crashing fertility is not just a phenomenon of the rich, developed world, or of countries with draconian controls. It is now happening at a breakneck pace even in the most unlikely places.

Bangladesh is among the world's poorest countries. Its girls are among the world's least educated and often marry when barely into their teens. Demographers used to argue that societies needed to modernize and prosper before they could put the brake on fertility. Such arguments fail in a country like Bangladesh, as does the notion that girls need to be educated and dissuaded from marrying early if they are not to have big families. Discredited, too, are the Malthusians' fears that given half a chance, the poor will breed their way to oblivion. So are the siren voices of those population planners who insist that such people must be frog-marched to contraception for their own good. It isn't so. None of it. All these things may help reduce fertility. Bangladesh shows none of them are necessary.

Bangladesh's women are having families half the size of their mothers. Out of choice. Back in 1999, Australian demographer Jack Caldwell noted that "among countries where there has not been a long-term coercive government family planning programme, Bangladesh is the poorest to have a total fertility rate under five births per woman." A decade later, it is the poorest with a fertility rate of three per woman. Something important has happened here. But what?

Family planning professionals claim Bangladesh is a benchmark success in their increasingly sophisticated efforts to promote contraception among poor women. After the failed efforts at a military-style rollout of contraceptives in Bangladesh in the early 1980s, the men with guns and snarling faces withdrew. A new generation of women from foreign population agencies chose Bangladesh as a good place to try persuasion with a smiling female face. They started by blitzing the Matlab district, southeast of Dhaka. At one stage, fifty thousand young, well-educated, and highly trained female family planning workers toured the countryside, visiting women in their homes every two weeks, offering them a wide variety of contraceptive methods. IUD uptake in particular was huge. Fertility went down by half in fifteen years. An influential World Bank study carried out in 1994 by John Cleland of the London School of Hygiene and Tropical Medicine concluded that the results in Matlab showed that family planning services could deliver falling fertility almost anywhere. The crucial thing was to get on and do it.

But how important exactly were all those health workers to the falling fertility? Some think their role has been dramatically overstated. During the 1980s, family planning took off right across Bangladesh, regardless of whether there was an invasion of health workers. The truth seems to be that Bangladeshi women wanted and needed contraception. They grabbed it. Of course, delivering contraceptives to women and offering them family planning advice both mattered. But the crucial fact was that Bangladesh women wanted and demanded them.

Caldwell sees the green revolution as critical to that decision. It reduced farmers' need for labor. Children, the traditional farm drudges, are no longer so useful. They get in the way. Farmers instead need cash to invest in fertilizer and seeds. So to be useful, the kids have to find work. That requires education. And educating children is expensive. Result: small families. And young women like Akhi, Aisha, and Miriam are sent to the cities to earn cash.

That makes sense. When I first went to Bangladesh in the 1980s, when fertility was still high, the countryside was green, pleasant, and still largely cashless. Rice was the only thing that mattered. There were few shops or roadside traders. Now traders are everywhere, and cash is vital to pay for fertilizer, schooling, TVs, doctors' bills, electricity supplies, and backyard tube wells. Meanwhile, Dhaka has turned from a sleepy old colonial capital into a teeming megacity.

But this reproductive revolution is not just about the economics of the green revolution. It is also about culture and the ambition of women. Mashuda Shefali Khatun, a campaigner at the Center for Women's Initiatives in Dhaka, told me, "Rural women in Bangladesh are proud of their small families. And they all want jobs now. It is partly economic need, but also because it increases their freedom. Even working in a sweatshop in Dhaka is freedom for them." As we spoke in Mohakhali, I noticed that Akhi wore a branded Adidas satchel over her sari. And hanging on a hook in the back of the room was Aisha's handbag, marked "Gucci." It was a fake, of course, but evidence of the new urban world.

. . .

Asian women are voting with their wombs. Family planning seems to take hold independently of nationhood or religion, regardless of national political backing, and often in spite of political opposition. Burma is nobody's idea of an open society. Its military junta refuses to let foreign journalists in or its people out. Western media are banned, and Internet access is tightly controlled. The government is opposed to family planning. It says it wants to double the country's population. But there is a strong feminist streak in Burma. Two decades ago, its people elected as prime minister Aung San Suu Kyi, although the generals have since prevented her from taking power. While the military may have a tight control on public life, its writ does not run in the home or bedroom. There, Burmese women are a law unto themselves. Whatever the generals might want, a lot of Burmese women don't marry—including 40 percent of those with a university education. They don't want babies either. Peter McDonald of the Australian National University says contraception is smuggled in from Thailand. The country's fertility rate is now reckoned at just 1.9 children per woman. The generals can do nothing about it.

And then there is Iran.

I first realized there was something different about Iran when reporting from the Cairo UN population conference in 1994. The mullahs called a rare press conference. Journalists showed up expecting them to attack family planning as a Western plot. There was even talk of a grand "faith coalition" with the Vatican. But as deputy health minister Mohamed Ali Tashkiri headed for the platform, I noticed the Iranian TV crew. All elbows and tripods, with heavy cameras on their shoulders, pushing their way to the front to ensure a prime spot for the Teheran evening news, they were dressed in full black burkas. The crew were all women. It was the same in the press room, with fetching Persian eyes barely visible behind veils as the female reporters filed copy from their laptops. Iran, I concluded, was special.

The women of some of Iran's Muslim neighbors still have lots of babies, averaging more than six apiece in Afghanistan and Yemen. But in the land of the ayatollahs, after three decades of fundamentalist rule, the average Iranian woman today has 1.7 children. The kind of male clerical dominance that can insist on a burka does not, it seems, preclude low fertility. Perhaps it ends up encouraging it.

Clerics took power in Iran in 1979, when Ayatollah Khomeini re-

placed the Shah of Iran after an Islamic revolution. The clerics immediately dismantled the Shah's modest and ineffective family planning
system. Soon they were at war, sending millions of young soldiers to
fight Iraq. Half a million never returned. Khomeini called for Iran's
women to provide a new generation of warriors. They obliged. Fertility
peaked at about seven children per woman, rising to nine in the countryside, says Amir Mehryar, who runs the Center for Population Research in Teheran and saw it all. Only after 1985, as the war ground to a
halt and census returns showed a population likely to double in twenty
years, did the clerics take fright and accept advice from their health professionals to allow family planning. Iran's women were, most probably,
exhausted anyway.

The 1990s brought electricity and piped water and schools to the
countryside. Revolutionary baby making gave way to consumerism as
millions bought TVs, refrigerators, and cars—as well as grabbing the
free condoms, pills, and IUDs doled out by government clinics. Those
years saw Iran experience "the largest and fastest fall in fertility ever recorded anywhere in the world," says Peter McDonald. Greater even than
the fall in post-Mao China.

Much later, in 2006, President Mahmoud Ahmadinejad tried to
wind back the clock by calling for women to return to their "main mission" of having babies. "Our country has a lot of capacity—even for 120
million people," he said. (Iran's present population is 67 million.) But
he was out of line with both women and his own administration, which
runs a condom factory west of Teheran. It produces 45 million sheaths
a year in thirty different shapes, colors, and flavors (pink and mint are
the favorites).

Many have seen the extraordinary collapse in Iranian fertility as a
triumph for westernization and a defeat for Islam. That is partly true. It
coincides with both a rise in consumerism and much greater educational
opportunities for women. But as in Burma, Bangladesh, and elsewhere,
the dynamic is less about westernization and more about the rising
power and influence of women.

Iran today is in a social ferment. We saw that in the disputes over
the outcome of the 2009 presidential elections, when the claimed victory by hard-liners was contested by reformers on the streets of Tehran.
But the ferment is at least as great between men and women. In the process, having babies seems to have fallen to the bottom of the list of pri-

orities for both sexes, says Soraya Tremayne, an Iranian-born academic now working at Oxford University. Iran is a country where women can buy chic Western fashion, but may get arrested if they walk to the store without their heads covered. Where women outnumber men in universities, but only 12 percent have jobs. Where a young female engineering graduate recently incinerated herself after returning home to be told that her father was away and she was to be controlled by her six-year-old brother.

In this battle, men rule the public sphere. They control marriage law. Judges still regularly prevent even middle-aged professional women from marrying the men of their choice, if their fathers disapprove. Women still have to bow before the traditions of arranged marriages, especially long-standing arrangements. Two octogenarian men recently married each other's fourteen-year-old daughters. Girls who step out of line are still sometimes killed. Men also interpret Shiite law to allow them to indulge in temporary marriages. These can take place anytime after puberty—which for a girl is deemed to be at eight years and nine months old.

But women are fighting back. They rule in the house and often within marriages. "Divorce rates are soaring because women get rid of husbands who beat them or take drugs or go to prison," says Tremayne. And they are insisting on control of their reproduction. Around three-quarters of married couples regularly use modern contraception, mostly the pill and condoms. This is the highest rate in the Muslim world. Women typically marry young, have a first child soon, and then start taking the pill. If they have a second child at all, it is an average of six years later. Rather like southern and eastern Europe, Iran seems headed for ultralow fertility, because of the clash between a patriarchal state and frustrated feminism.

. . .

Bangladesh and Iran are far from alone in the Muslim world in seeing crashing fertility. Morocco (with a fertility rate of 2.6 children per woman), Tunisia and Iran (both 1.7), and Turkey (2.2) have all made the same journey. In the twenty-first century, few clerics have either the inclination or the ability to prevent their flocks from adopting modern birth control.

The Vatican has fared little better. It long since gave up seriously trying to impose its writ in European bedrooms. But it has failed, too, in most of Latin America. In Brazil, home to more Catholics than any other country, the church has for decades blocked the establishment of a state family planning service. Even so, 40 percent of married women of reproductive age go to private clinics to be sterilized. Perhaps their choice of contraception is a compromise with the church. You have to confess only once to being sterilized. But when a *New York Times* reporter visited a health clinic in a poor part of Rio de Janeiro and asked the gynecologist whether women ever brought up religious objections when he discussed birth control with them, he replied, "Well, I do remember a case last year. I think she was a Jehovah's Witness."

Brazilian fertility has been halved in twenty-five years, to 2.2 children per woman. Mexico, El Salvador, Ecuador, Nicaragua, Colombia, Venezuela, and Peru have all seen their fertility rates fall from more than six children per woman to less than three. Some slowpokes are still above three: Guatemala, Bolivia, Honduras, and Paraguay. But even they have halved their rates. Chile and Costa Rica are down to two, and Argentina to 2.4. On the islands of the Caribbean, most countries are already below replacement level. In Cuba, where Catholicism has been replaced by communism, fertility has fallen to 1.6. The only serious holdout is Haiti, where Voodoo has equal status with Catholicism. Its fertility remains at 4.8.

. . .

Across the world today, for the first time in history, most sex acts are prevented from producing children. Like the mobile phone and television, contraception is a near-universal technology. Only the most bizarre and patriarchal regimes try to withhold it. People have sex somewhere between 100 million and 200 million times every day, depending on whom you believe and how you define it. This results in something like 900,000 conceptions. Some 100,000 of those will be deliberately aborted, and some 370,000 will result in live births.

In China, over 80 percent of couples use contraception most of the time. In Latin America, 70 percent do. But across sub-Saharan Africa, only around a quarter of couples take precautions using modern methods, though that is still more than double the figure for their parents. In

Muslim Afghanistan, it is still under 5 percent. An estimated 76 million unplanned pregnancies still happen round the world each year. About half end in abortions and the other half in live births. So of the world's 135 million or so births each year, around 100 million are planned and 35 million not. Of the latter, many may simply have happened earlier than intended and therefore might not be births that never would have happened. But that is not clear.

Contraception is overwhelmingly women's business. Round the world, men take responsibility for contraception less than a tenth of the time, with most using a condom. The other 90 percent is down to women. In rich countries, women are much more likely to use the pill. In poor countries, they predominantly use methods that are one-off, such as sterilization, or require only occasional attention. This is partly because health workers push these methods, believing women cannot be "trusted" to take a pill; partly because distribution of products needed regularly can be unreliable; and partly because these methods prevent men from being involved, or sometimes even knowing what is going on.

More than half of all the world's IUDs are implanted in China. In many countries, especially those where the Catholic Church holds sway, female sterilization is very popular. Almost half of all women in Puerto Rico and the Dominican Republic have been sterilized, and more than a third in Brazil, China, and India, where, thanks to intense government promotion, female sterilization makes up four-fifths of all contraception use. Injectable hormone contraceptives, such as Depo Provera, are still controversial because of possible side effects. But where they are promoted by governments—for instance, in South Africa, Indonesia, Haiti, Mongolia, and Kenya—they pick up between 20 and 40 percent of the market.

The condom is more popular in the rich world than elsewhere. This is partly because men there are more willing to take responsibility for stopping babies, partly to ward off AIDS, and partly because some women in rich countries are losing confidence in the pill. Spaniards are now the second biggest condoms users at 34 percent. But the Japanese, once the acknowledged condom kings with 80 percent market penetration, have halved use in twenty-five years. Vaginal barriers are losing popularity, except in Japan, where they are replacing the condom.

Male willingness to take responsibility may also lie behind the otherwise unexpected popularity in eastern Europe of the most traditional of methods of preventing babies. Withdrawal is still tops in Romania, Slovakia, and Bosnia. In Albania, if you believe the figures, 90 percent of intercourse between reproductive couples ends in extracorporeal ejaculation. But there are limits to what men will willingly do. Male sterilization, which is much easier and generally safer than female sterilization, is rare in most countries. A generation ago, more men than women were sterilized in India, Bangladesh, and South Korea. No longer. Anglo-Saxons seem most amenable to the snip today. New Zealand has the world's highest vasectomy rate, with nearly one in five married men having gone under the knife. Britain is close behind. Nobody else is anywhere near.

. . .

The feminization of both contraception and population policy is one of the major changes of the past two decades and underpins the rapid worldwide fall in fertility. The 1994 UN population conference in Cairo, where I watched burka-wearing TV crews, was the occasion when the language and personnel of family planning finally passed from men to women. Both the male population controllers, heirs to Rockefeller and the other postwar pioneers, and their foes among the clerics had to make way for female reproductive rights advocates. Headlines from the event reflected this. "Sisters are doing it for themselves," "Women take the lead in Cairo battle of sexes," and "Women hold key to population curb" were typical. Benazir Bhutto and Jane Fonda turned up for the show.

The UN Fund for Population Activities was by then in the hands of a female gynecologist, Nafis Sadik, rather than Rockefeller's protégé, Rafael Salas, who had recently died in office after eighteen years. Sadik was scathing in her condemnation of the past male hegemony. "If we had paid more attention to empowering women thirty years ago, and had listened to their needs, we might well have been ahead of the game as far as population numbers are concerned. If you really looked after women's needs and women's health, everything else would take care of itself. Not allowing them to have the capacity to take decisions for themselves is really the main obstacle to population control."

Coercion, she said, was not just unpleasant and unfair, it was also counterproductive. Fertility had usually gone down furthest where reproductive rights ruled, and least where contraception was imposed. Most women in most places now wanted to reduce their fertility. The worst way of encouraging them was to have men hectoring them to do it.

And she has been proved right. Since Cairo, dozens of countries have seen most women adopting contraception for the first time. Rich or poor, socialist or capitalist, patriarchal or not, educated or not, urban or rural, Muslim or Catholic, secular or devout, with tough government birth control policies or none, most countries are experiencing a reproductive revolution. However they do it, and whatever we think about it, the "population bomb" is being defused. By women. Because they want to.

CHAPTER 14

Sex and the City

Carrie was a partying columnist who wrote about love and sex. Miranda was a Harvard-educated lawyer who planned her life on notepads and had a pet househusband. Samantha loved sex and thought like a man. Charlotte converted to Judaism in order to marry her divorce lawyer. What did these New York women have in common that no previous generation of women would have been able to claim? The answer, of course, is that the four lead characters of the TV series and movie *Sex and the City* were well into their thirties and, for most of the show's life, had no babies. And nobody saw this as unusual, let alone antisocial.

In most times, in virtually all countries, they would have had a brood of children by that age. Barring a few career spinsters and largely closeted lesbians, every woman did. The social pressures for women to be fertile were so strong that few resisted, or were allowed to resist. In parts of the world today, that is still the case—in some Middle Eastern countries and parts of sub-Saharan Africa, for instance. But in most places there has been a revolution in the way women see themselves, and increasingly in how society sees them.

I have only rarely watched *Sex and the City*. I'm not in the right demographic. But traveling the world in the past decade, I know its characters and lifestyles as an ever-present cultural riff. They are on TV from Lagos to Shanghai, Hong Kong to Buenos Aires, and Nairobi to Jakarta. Russia created its own Moscow version, *The Balzac Age*, subtitled *All Men Are Bastards*.

People comment on how the original differs from their own societies. Samantha would have had a sex change by now in Mumbai, says *Times of India* columnist Namita Devidayal. But it is the similarities that are most striking. Chen Chang, a thirty-one-year-old film producer in Beijing, is one of millions of Chinese women who watched *Sex and the City*. She told the BBC, "They have all the same problems as we do, having a career and a man, the right man, whether to have children, divorce, how to have good sex." Her friends are in advertising and fashion, drink in fashionable bars, shop for imported clothes, and travel abroad. It's quite a change. "My grandmother was sold to a wealthy silk trader as his fourth concubine. She had no education, no opportunities, and she had bound feet, so she couldn't even leave the house. And she committed suicide after failing to give birth to a son."

Handbooks for being a single girl, like Sunny Singh's *Single in the City*, are best sellers. Frothy columnists talk of a "new girl order" as "single young females take over the globe." Besides *Sex and the City*, Maryland psychologist Linda Berg-Cross points to the popularity of *Ally McBeal* and *Bridget Jones*, and notes that "all the principal professional women on *The West Wing* were single, save the first lady (who lost her medical license standing by her man)." This all reflects a real social change, says Berg-Cross: "The globalization of elite, single professional women is the first new global sociological phenomenon of the twenty-first century."

But it is not just elite professionals who are joining in. The world over, women are staying single through their twenties and beyond. In 1960 two-thirds of American women in their early twenties were already married; today the figure is less than a quarter. In Hungary, 30 percent of women in their thirties are single, compared to 6 percent of their mothers' generation. As recently as 1980, a UN study found that nuptials were "near universal" across Asia, and half of Asian women were married by the age of eighteen. No more. In Japan, half of all thirty-year-old women today are unmarried. In South Korea, where there is an increasingly entrenched celibacy among educated women, the figure is 40 percent. Their mothers finished school, married, and had kids. The new generation go to college, most likely university, and then get jobs. In the city. In Bangkok, a fifth of all women are single at forty-five. Manila, Singapore, Hong Kong, and the Chinese population of Kuala Lumpur are not far behind.

Australian Bernard Salt, author of a book called *Man Drought*, says Aussie women go to work in offices in cities, while men head for mines and farms and industrial zones. "A generation ago, women were content to live in rural Australia and marry the guy next door." They had their first child at twenty-three years old on average. "Now they travel into cities, get educated and have careers." They don't even think about "settling down" until their thirties.

Unremarked, it is women who are leading the urbanization of the planet. Whether sweatshop workers in Dhaka, bar girls in Bangkok, office workers in Shanghai, students in Delhi, or maids in Caracas, there are more young women than men in almost every city in the world. Academics debate whether they are there for the jobs or the men (there are more rich, educated men in cities, but more women of all sorts). But whatever the motive, they are there in the bars, shoe shops, gyms, and clubs. With jobs but no dependents, they have money and a lifestyle that in Japan gets them called *wagamama* ("selfish") or "parasite singles." No matter; it's better than changing diapers.

. . .

Future historians are likely to record two great social trends in the last half of the twentieth century—the dramatic decline in fertility and the transformation of the role of women in society. These two events are clearly linked. But which came first? Some say liberation allowed women to make choices about their sexuality, their partners, and the size of their families. Equally, however, it has been the dramatic improvement in the survival rate of infants that for the first time has freed women from the social obligation for a lifetime of producing and rearing babies—and given them the chance to assert their role in society. Scottish sociologist John MacInnes calls this the "reproductive revolution" and sees it as the greatest and most long-lasting revolution of the twentieth century.

Until the eighteenth century, half of all offspring died before entering their fertile years, and many more died before they completed them. Most women spent almost all their (often rather short) adult lives bearing and rearing children. Now most children reach adulthood, and most adult women live out their fertile life span. Women can have far fewer children than they once did. And they have much longer lives in which to do it. Bearing children typically take up 10–15 percent of their adult lives.

Women have grabbed the chance created by that change. While having children remains very important to most women's lives, it is no longer the only thing, or even the main thing, they do. They cease to wield their power only within the home. Now they are out of the front door. The results of this revolution are evident everywhere—not least in lots of thirty-something career women still being single, free, and sexy in the cities across the world, knowing there is plenty of time for having a baby or two if they choose. Knowing, too, that society will be quite relaxed if they choose not to.

Across the rich world, and much of the poorer world too, women outnumber men on university campuses and dominate entry to professions like medicine, media, and the law. They run the farms and even the governments, sometimes. The reproductive revolution has created a feminist revolution that still has a long way to go. But it has already changed the world.

British demographer Tim Dyson highlights the extent of the change even in rural India. India celebrated its billionth citizen in May 2000. It still adds about nineteen million people to its population each year—a quarter of the global growth. But fertility is falling fast, now averaging 2.8 children per woman. In southern India, it is close to two. Dyson remembers that back in the 1960s, sociologists compared women's lives in India and the United States. "In India they married at seventeen, had seven kids, the last one at forty-three years old, and died, typically at forty-six. In America, women typically married at eighteen, had two kids quickly and then more than forty years of life after childbearing. Now Indian women are grabbing that life too. Getting married and having children are simply not as important as they used to be. Sterilization is the main form of contraception in India, and the average age of sterilization is twenty-six years old." A recent study in Andhra Pradesh found that women married in the 1990s were now being sterilized, on average, five years after marriage, when still in their early twenties. This revolution has transformed the role of women in India. Go to any community now, and it is the women who turn up at the farm gate to greet visitors, who are the doctors and academics and who run the NGOs (nongovernmental organizations), says Dyson. "The women there are just more businesslike and practical than the men. There is a huge reduction in gender differentiation going on in India."

This is a staggering change. For thousands of years, male-dominated societies were nearly universal. Men ruled the world. MacInnes argues that this was no accident. Patriarchy was regarded as essential to ensure that women produced the next generation. It was deeply engrained and tenaciously defended by men. The regulation of child production was done "through church and state, through the norms surrounding sexual activity and sex roles, illegitimacy, cohabitation and marriage, through family and kinship obligations, and property law." It sometimes required the sexes to be segregated. It demonized illegitimate children. It ostracized any form of homosexuality. Lesbianism was often simply decreed not to exist. In Asia, patriarchy involved arranged marriages, often of very young girls, and brutal sanctions against female adulterers or girls who would not accept their lot.

The reproductive revolution kicked away this system of patriarchy, because it was simply no longer needed to sustain populations. Women have always wanted equal rights. Feminism is not a new idea. And some individuals have always broken free. But for most women, the reproductive revolution has taken feminism "from the realm of utopia to practical possibility," says MacInnes.

I like this argument. It explains, in a way that nothing else does, why fertility has fallen almost everywhere round the world. Women are giving up on constant childbirth for the simple reason that for the first time in history, they can. It is no coincidence, surely, that patriarchy is collapsing at the same time. This is not the slow diffusion of a new idea or the success of a cold-war policy cooked up in Washington, nor a mechanistic response to aid workers handing out condoms. This is the breaking of a logjam—the logjam of a patriarchy that has suddenly lost its purpose.

. . .

So where is this taking us? Some anticipate population stability, though increasingly, population decline looks likely. But first, what is required for population stability? Through most of history, women have had between five and eight live births each. They needed to, in order to maintain populations in the face of high death rates among children. Today they average 2.6 births globally. But how many children should women have to secure future generations? The simple answer would be two:

one to replace each parent. But it is always going to be more than that. Not every child grows up, ready and willing to produce the next generation. Through most of history, more than half of children were cut down by diseases like measles, malaria, dysentery, and smallpox or some other mishap. A study of the U.S. census of 1821 showed that eleven out of every twenty girls survived to middle age, and one of those eleven remained single, suggesting that couples needed to have four children to maintain population. That was probably as good as it got back then.

Today the great majority of babies reach adulthood. So what is the new "replacement rate" for fertility? It is not exactly two, because nature delivers about 105 boys for every hundred girls. This compensates for a generally lower life expectancy among males. So for a hundred women to produce another hundred women to carry on the line, they also produce about 105 boys, creating a replacement fertility rate of 2.05. Since not all children grow up even today, modern demographers plump for 2.1 children for every woman as a rule-of-thumb replacement level.

That's about right for rich, developed countries. But it gets universally quoted, even by demographers, as if it were a hard and fast rule, even for countries where many more children die young. This is sloppy. In 2003, demographers at Princeton University's Office of Population Research, headed by Tom Espenshade, reworked the figures to come up with a replacement fertility level for each country, allowing for different survival rates. They found a huge range, from less than 2.1 babies per woman to almost 3.5.

Nine countries have real replacement rates above 3. Eight are in Africa; the ninth is Afghanistan. The reason is simple: large numbers of girls die before they have a chance to reproduce. In the southern African countries of Swaziland and Botswana, the main reason is the high death toll among children from AIDS. They catch the virus at birth through their mother's blood. In tiny, landlocked Swaziland, a staggering 40 percent of the population has HIV. Life expectancy is thirty-two years. That is half the global average. Espenshade calculates that Swazi women each need to produce 3.35 children to maintain the population in the long term. The last estimate of the country's actual fertility was 3.2. Its population is falling. Botswana and South Africa have current fertility of around 2.5, rather lower than their AIDS-influenced replacement levels of 3.0 and 2.6, respectively. Both countries are on course for a falling population if HIV persists at its current level.

According to Espenshade, global replacement fertility level—the average number of babies per woman required across the world in the long term to maintain the current population—is 2.3. How far are we from this? The global fertility rate has been falling fast. From between 5 and 6 in the early 1950s, it fell by the late 1970s to 3.9. By 2000 it had come down to 2.8, and by 2008 it was at 2.6. Already, more than sixty countries—containing close to half of the world's population—have fertility rates below national replacement levels. The club now includes most of the Caribbean islands, Japan, South Korea, China, Thailand, Sri Lanka, Iran, Turkey, Vietnam, Brazil, Algeria, Kazakhstan, and Tunisia. Within twenty years, demographic giants like Indonesia, Mexico, and India will in all probability have below-replacement fertility.

At the current rate of decline, the world's fertility rate will be below replacement level soon after 2020. It could happen sooner if the world's potential parents react to the current economic downturn, as Europe did during the depression of the 1930s, by having fewer babies. In any event, on current trends, the world's population is primed to start falling for probably the first time since the Black Death in the fourteenth century.

CHAPTER 15

Singapore Sling

The feminist revolution is a long way from being completed. In much of the world, men still want to control their womenfolk. But ultimately, we may face a choice: complete the revolution or sink into a mire of ultralow fertility. Why? Because the evidence from dozens of countries that have passed through the replacement fertility barrier is that without a new attitude from men, the state, religious authorities, and employers, women are going to give up on childbirth altogether. Given a choice between having children and having a life, they will choose a life.

For Singapore's current prime minister, Lee Hsien Loong, this conumdrum is creating a national emergency. The native population of Singapore could be cut in half by midcentury, making the natives a minority in their own land. The women of Singapore are on "baby strike," Lee said in his National Day address in August 2008. They are producing fewer than forty thousand babies a year. A generation more of that, and Singapore will have reached a "point of no return." It would be the end for a shining city-state less than half a century old.

The papers were full of Lee's National Day address—not least because he boasted that he lived the enlightened life he wanted his fellow men to adopt. Being a father of four and well able to change a diaper, he plays the new man, a twenty-first-century father of the nation. "If husbands leave everything to the wives, or the women are forced to choose between working and having babies, they are going to go on baby strike." Men should do more, he said. (He didn't mention that behind the scenes, employers had vetoed his suggestions that the state should give fathers a helping hand by enforcing paternity leave for new fathers. They said it would disrupt the economy too much. In Singapore nothing, not even national survival, is allowed to disrupt the economy.)

The *Straits Times* was full of diagnosis of Singapore's demographic crisis. Many contributors blamed the birth dearth on the very single-mindedness that has made Singapore one of the economic success stories of the late twentieth century. Singapore is today the richest of the Asian tiger economies. But people have been so busy getting rich that they have omitted to produce the next generation.

"Our pursuit of economic success over the past 40 years has resulted in a new breed of young Singaporeans who hold different values about life," said one fairly representative columnist. "They are individualistic, mobile and materialistic. They do not want to sacrifice their personal freedom for family life. They socialise more, but they fight shy of commitment." Another warned that "children who grow up in the company of domestic help and spend all their time at tuition or enrichment classes may not want children of their own."

Some blamed women for the birth dearth. The shirtsleeved prime minister said men's failure to change diapers threatened national security. "If I can do it, anybody can do it." Others blamed the country's string of elite single-sex schools, dating back to the time of the city's founder, Sir Stamford Raffles. They have produced well-educated but "uncultured and uncouth" boys and "backstabbing" girls. They harbor "liberal intellectuals" who encourage homosexuality.

Nobody mentioned the man who arguably landed the country in this pickle: the current prime minister's father, former prime minister Lee Kwan Yu, still alive at eighty-four.

The city-state of Singapore is on an island less than half the size of London and has a population of four million. It is a society built on control of every aspect of life. If China is made up of communists running a capitalist state, Singapore is made up of capitalists running a socialist, or at least a nanny, state. Its efforts to control its population go back to the origins of the country, following the breakup of the British Empire in Southeast Asia.

In the mid-1950s, an average Singaporean woman had six children. But with death rates in freefall and migrants rushing in, the country's population almost doubled in the thirteen years between 1957 and 1970.

This was too much. In 1966 Lee Kuan Yew created the Singapore Family Planning and Population Board "to initiate and undertake population control programs." He turned the full force of a formidable government propaganda machine toward reducing family size, churning out posters with slogans like "Stop at Two," "Two is Enough," and "Take Your Time to Say 'Yes.'"

It worked. Lee's people readily complied with the new policy. Fertility fell. On advice from Western demographers, Lee assumed that fertility would stabilize around replacement level. But the policy was too successful. The "Stop at Two" adverts continued to appear until 1982, by which time a lot of people were stopping at one and fertility was down to 1.7 babies per woman. Only then did Lee call a halt to the propaganda.

Alarmed at the success of his old policies, he backtracked and started promoting larger families instead. Well, for some. His policies for promoting baby making were avowedly eugenic. The 1980 Singapore census had revealed that fertility had fallen furthest among the best educated, while the less educated were still "overproducing." Lee got to the nub at the 1983 National Day rally: "If you don't include women graduates in your breeding pool and leave them on the shelf, you end up a more stupid society.... There will be less bright people to support dumb people in the next generation. That's a problem."

So Lee decided to encourage marriage and baby making specifically among graduates. In 1984 the new Social Development Unit began offering tax relief on babies and preferential access to the best schools—but only to the educated. The lower orders were left out. Instead they were offered cash incentives to be sterilized. Women without educational qualifications could pocket ten thousand dollars if they agreed to the procedure after their first or second child.

Some said this policy grew more from racial concerns than from any fears about the dumbing down of the nation—that it was designed to bolster the continued domination of ethnic Chinese against the more fertile Malay minority. Certainly, race has always been a hidden dimension in Singapore's politics. The state itself seceded from Malaysia in the early 1960s after disputes between Chinese and Malays spilled over into race riots. Equally certainly, the Chinese community is better educated and has fewer children.

Whatever the prime minister's intentions, the fertility of all races and classes continued to fall—and in 1987, Lee decided he cared as much about quantity as quality and extended the pronatalist strategy to all. The government set up a second matchmaking service specially to target nongraduates. They were hectored with slogans like "Make room for love in your life" and "Life's fun when you're a dad and mum" and the less-than-aphrodisiac "Have three or more children if you can afford it." There was relief when a brief baby boom took hold in the late 1980s. But it didn't last long. Soon the birth dearth had escalated. In 2000 Lee's successor, Goh Chok Tong, introduced a cash "baby bonus." By 2004 Lee Jr. was in charge, and he extended maternity leave to three months. The state waded in deeper, funding charities like the "profertility" I Love Children, the Working Mothers Forum, and Parenting@work.

None of these efforts have met with any success. Two decades after Singapore imposed the world's longest-lasting pronatalist policies, its fertility rates are even lower than they were at the start. Fertility was 1.1 children per woman—the world's lowest—in 2009, the fifth year running that the rate had been below 1.3. Yap Mui Teng, a government adviser and social scientist at the University of Singapore, explained the problem to me as we awaited Lee's national TV address: "When they were trying to lower fertility, the goals of society and individuals were the same, so change was swift. But now individuals and society are pulling in different directions—and individuals are winning."

Singapore's ethnic Chinese have been the most resistant to the new message. Their fertility has been consistently below 1.1 in recent years, just half the level of ethnic Malays. Ironically, Singapore's industrious Chinese have aped mainland China's one-child policy by choice, when their own government wants them to have two or three. Only in Dragon Years like 1976, 1988, and 2000—when children are traditionally expected to be endowed with power, luck, and charisma—do the Chinese stir themselves to conceive babies. But don't expect too much of a baby bounce in 2012, for 2000 produced only a minor baby boom.

Singapore is the only country on earth where the government plays matchmaker. It believes that part of the problem is that its people are too shy and too preoccupied with work to pair up. So it runs online services like Lovebyte, which also offers counselor Dr. Love and tips on personal hygiene, and its new Romancing Singapore Web site. The latest idea is "relationship" classes for higher education students. At

Singapore Polytechnic, "Love relations for life—a journey of romance, love and sexuality" is one of the top ten most popular courses, with seven hundred students. More primly, the National University offers a module on "Dynamics of Interpersonal Effectiveness." It features "self-awareness, others-awareness, emotional intelligence, social cognition and attitude formation," along with the teasing promise of "practical skills using experiential learning methods."

"People do not have time to socialize, and they also do not know how to socialize," said one teacher at the University of Singapore. Students have been queuing up for the lessons, while insisting that they join because they are almost guaranteed A grades. "It's easy to score," sniggered one student. Cynicism about government efforts to encourage love, sex, and babies is widespread. An old joke is that the Social Development Unit's initials actually stand for "Single, desperate, and unwanted." Akasha, a young public relations executive, told me that "all this propaganda ends up turning people off. They say the government shouldn't interfere."

Can private matchmakers score better? They are certainly milking the government's enthusiasm to find out. The government funds Heart2Heart Connect, gomoviedate.com, and Club2040, which claims to have set the world record for the most people speed dating in one venue at the same time. During my short stay—had I been a little younger and single—I could have joined training sessions in "16 secret moves that get ladies interested in you on dates," baking classes, tea dancing, a "professionals networking nite" at the Lovebyte café, board games, a "boot camp for men" to discuss "girlfriend-getting secrets," an event to "discover your love language," and an all-night car rally aimed to "turn on the lights of love" as "participants get to mingle in small groups of three cars each"—all courtesy of government matchmakers.

It sounds like a mixture of Orwellian patriarchy and Gilbert and Sullivan farce. In other countries, such propaganda might be dismissed as a heavy-handed attempt at helping a few shy youngsters through a difficult phase of their life. Here it is about national survival.

Singapore's Chinese kids are not getting hitched, says Yap. It is more than a failure of social skills. Among men, the ill-educated marry the least, she says. More than a quarter of those without secondary education are unmarried at their fortieth birthday, compared with only 13 percent of graduates. But among women, the graph is in the opposite direction. Only 10 percent of women without secondary schooling are single at age forty, but among graduates, the figure is 30 percent.

The story seems to be this. Educated women with careers are extremely wary of marriage, which might interrupt their ambitions. And they certainly don't want to hitch up with men less well educated than them, at least not for marriage. Well-educated men and less-educated women seem to shack up quite successfully. But less-educated men don't stand a chance. Many of them are now looking abroad to find women, whether potential wives in the villages of mainland China and Vietnam or prostitutes in the brothels of Thailand. Meanwhile, Singapore keeps going only thanks to huge influxes of foreigners. The TV broadcast of Lee's national address in English was delayed for twenty-four hours so as not to clash with the live broadcast of Singapore's only medal hopefuls, the women's table tennis team, in action at the Beijing Olympics. They only got silver. But as many noted, they were not native Singaporeans, but Chinese migrants. In fact, half the national Olympic team turned out to be nonnatives.

What about children? Singapore does still have some. I went to see the country's top pediatrician, Daniel Goh, in his office at the National University Hospital. His waiting room was the one place in Singapore where I saw children in any numbers. It came as a huge shock. I had almost forgotten that they existed. Their numbers are so small, and most of the time they are either in school, attending cramming classes, or sweating out homework in apartment blocks across the city.

The country's breakneck economic growth has been built on education. According to demographer Wolfgang Lutz, Singapore "has probably experienced the most rapid educational expansion in recent history." While most Singaporeans over age fifty have never attended school, and many are illiterate, more than a half of those in their twenties—among the highest figures in the world—have studied for a degree. For parents, the lesson is personal and obvious: educational qualifications are their children's route to prosperity. But at what cost? The pressure on parents and children alike is intense, and the investment of time, energy, and cash often extreme.

To get good grades, extensive coaching is almost compulsory. So much so that children feel left out if they don't get it, said Goh. He spoke from personal experience. "My wife and I fought against the cramming system with our own three boys," he said. "They are aged six, ten, and eleven. We decided not to buy them extra tuition. We wanted them to have time to grow as people. But in the end we felt negligent for not doing so, because everybody else did."

The Singapore education system is very academic, and children are constantly evaluated and rigorously segregated according to their class grade. "We held out till our oldest was nine and he was falling behind in his language classes," said Goh. "Having given in, we had to do the same for the other two children. In the end, it creates more stress not having tuition than having it. But they are losing their freer, happier life."

The posters may say "Life's fun when you're a dad and mum," but it doesn't feel much like it for parents who are as pressured as their children by the hothouse educational atmosphere. The prospect of parenthood is daunting. So even as fewer and fewer Singaporeans marry, fewer of those who do marry are having children. And more and more of them stop at one. Goh told me, "In my job, I see increasing numbers of married couples who are not having children. Before, that was socially unacceptable. Now it is quite normal."

The trouble is that Singapore's economic and social policies are in conflict. A quarter of mothers say they would have more children if the government organized better child care and more flexible working hours. But employers won't play ball, and the government won't force their hand. Says Lutz, "In Singapore, women work an average of fifty-three hours a week. Of course they are not going to have children. They don't have time."

Singapore seems to be a society falling out of love with children and out of the habit of producing them. Gavin Jones, the country's leading demographer, says social norms are changing. "You see it in small things. There has been a long series of complaints in the *Straits Times* recently about how parents use strollers as battering rams in shopping centers. One letter-writer said, 'Parents should leave their children at home.' Children just seem to be unacceptable members of society for some people. It's quite alarming. I was one of five children, and my wife is one of three, so we like children. It is normal for us to have children round the house. But I can imagine that in future small families will breed small families. People will get out of the habit." Many in their thirties, working late at the office and then unwinding in the city's bars and clubs and restaurants each night, already have.

Weeks after Lee's new initiatives in 2008 to boost baby making came the credit crunch. One of the first stories on the topic in the Singapore press was headlined "No money, no honey." Thousands of marriages were being put off because couples no longer had the cash to fund the nuptials. Even hot dates were on hold. Reuters called the receptionist at Truelove International Matchmaker. "Business has been very badly hit by the crisis," she said. "In the past, I would get around twenty calls a day. Now there are hardly any calls." Whether because people are too busy or too poor, even getting a date seems to be a problem in Asia's economic success story.

. . .

Singapore shows, in an exaggerated form, many of the stresses now emerging in low-fertility countries round the world, from Japan to Italy, Korea to Germany, and Canada to Ukraine—stresses that seem self-perpetuating and increasingly ingrained.

Like Singapore, many governments have tried to fight off low fertility with cash handouts to couples who have babies. Russia offers eleven thousand euros for second and subsequent children. Australia launched a baby bonus in 2004 under the slogan "one for Mum, one for Dad, and one for the country." Britain introduced a "baby bond," though it pays out only when the child reaches eighteen.

Governments claim that some of these payments have triggered a "baby bounce." Spain's 2,500-euro bonus coincided with a recovery of its fertility from 1.2 to 1.4 children per woman. But most analysts say couples are moving up plans for a baby rather than increasing their intended family size. The Australian government's chief adviser on demography, Peter McDonald, says, "Our bonus has about the same value as a wide-screen TV. People like it, but it doesn't change their decisions about having a family." German chancellor Helmut Kohl introduced tax incentives for babies way back in 1980, when fertility there was 1.41. He promised the incentives would add 200,000 births a year. They didn't, and Germany's fertility rate is today unmoved.

What couples really need—even those with cooperative diaper-wielding husbands—is practical help like flexible working hours, maternity and paternity leave, and child-care provision. Those are the things that have made a difference in northern Europe. It is no accident that fertility rates in Europe are highest where welfare systems work best. But they are things that many more patriarchal states and employers are reluctant to provide.

An anomaly here is the United States. It has managed to keep fertility rates as high as France or Scandinavia, but without their welfare provisions. This is partly because of the fertility boost from newly arrived Hispanic women, who initially average more than three children each. But even among the majority, U.S. fertility is at the Scandinavian rather than Italian level, according to Carl Haub of the Population Reference Bureau in Washington, D.C. The way the jobs market works is critical, he says. American women can't take a year of paid maternity leave, but they can quit their job and expect to resume their career later with minimal fuss. Equally important, he says, American men are much better fathers and husbands than their southern European counterparts. "In order to promote fertility," says Haub, "your society needs to be generous or flexible. The U.S. isn't very generous, but it is flexible. Italy isn't generous and it's not flexible."

Where do families and fertility go from here? The world as a whole could fall into the Mediterranean low-fertility trap, creating societies where children have little or no place. That could happen if men find that even a world of disposable diapers, pre-prepared food, and baby alarms has not quite converted them into new men willing to take their turn holding the baby. Especially if governments and employers won't play ball either. But the world could complete the feminization process begun by the reproductive revolution. It could allow women (and men, too) the opportunity both to have a family and to take their place in wider society.

These are urgent issues not just for rich countries, but also for those increasing numbers of developing and downright poor countries with low fertility. As women break their chains, will these countries follow the Scandinavian or Mediterranean or American model? Australian demographer Jack Caldwell warns that right now, "the Mediterranean patriarchal model is far more common in the world." There are more macho men than helpful husbands, more obstructive patriarchies than welfare states.

As women grab their new freedoms, the most pressing need is to instill new responsibilities in men. "If we are ever going to raise fertility back to anything like replacement level, then the key is changing men," says British demographer Tim Dyson. "Nothing else will work. Unless there is a renegotiation of gender roles, then below-replacement fertility will eventually be a fact everywhere." Women have become more like men, he says; now men must become more like women.

CHAPTER 16

Missing Girls

The reproductive revolution is creating some unexpected offspring. When couples are planning to have only one, two, or at most three children, they are much less keen to let fate decide what children they get—especially when technology can be deployed to ensure the "right" answer. Choosing the sex of your future child is now a way of life for many women in some countries. And the effects on societies can be profound.

In August 2008 the local newspapers in Mumbai were full of stories about maternity clinics being raided across the city on the orders of the Supreme Court in Delhi. These were not backstreet clinics, but the plusher, air-conditioned premises with the newest equipment and the smartest nurses, located in the most sought-after suburbs—clinics like the Malpani Infertility Clinic in swish Colaba, whose Web site advertises its services to British women as well as locals.

The government inspectors conducting the raids were seeking out ultrasound equipment and disabling it. Dr. Aniruddha Malpani and Dr. Anjali Malpani had three scanners that the inspectors claimed were being operated illegally. Later, the inspectors sealed fifteen more machines in the Chembur, Govandi, and South Mumbai districts. They spent a week in the city, looking for breaches of the 1994 Pre-natal Diagnostic Techniques (Regulation and Prevention of Misuse) Act, which requires that all ultrasound equipment be registered and every scan recorded.

The doctors claimed the infringements were technicalities. The inspectors said the lack of paperwork meant that patients who had been scanned couldn't be checked. The ultrasound gave pregnant women a sight of the fetuses inside their wombs, and the inspectors believed that rather than looking for congenital abnormalities, the women were searching for penises. And if they couldn't find one, they were signing up for an abortion. They don't want girls; they want only boys.

Hindu families feel they have to have a son to carry on the line, and traditionally a son has to light the funeral pyre when his parents die. Girls are often seen as a burden. They require dowries. And dowries are not fading away as India modernizes. They are becoming more elaborate, more status-tied, and more expensive. Despite being in short supply, girls are more and more expensive to marry off.

Sex-selective abortion is illegal in India, as in most countries. But it is a billion-rupee business in Mumbai. It happens enough to change the sex ratio of births registered in the city. In 2007, 7,400 more boys were born than girls in Mumbai. Nature delivers a 5 percent surplus of boys, but this excess amounted to almost 9 percent—suggesting that about 3,500 girl fetuses that should have been born in Mumbai in 2007 were not.

Indian authorities like to claim that aborting female fetuses is a cultural throwback, and that modernizing India is putting such archaic ways behind it. During the 2008 raids, the *Times of India* quoted a city official blaming its slum dwellers: "It is a fact that urban slums have the worst child sex ratio. Mumbai is no different." This is not a fact at all. Sex-selective abortion is spreading not in the slums, but among the middle classes, in exclusive suburbs like Colaba.

There has been a hoo-ha about sex-selective abortion in Mumbai primarily because of one man—a British-trained advertising executive who set up an NGO called Population First. It campaigns to shut down the miscreant clinics. I met Bobby Sista at his tiny office above a jewelry shop near the Standard Chartered Bank on Mahatma Gandhi Road. He told me, "When we saw the 2001 census results, we saw that Mumbai was one of the worst cities, and the worst sex ratios are in the most affluent areas. We started probing the clinics and found that most were not keeping proper records. It was obvious what was going on. But because it was an affliction of affluence—only the middle classes can afford the scans—the government was reluctant to act. So we set out to expose it."

Dozens of clinics in Mumbai are involved, he said. "One of the doctors nabbed in this week's swoop is rumored to practice sex-selective abortions in a big way. Most of the doctors are powerful, and they don't care about the law." This is a widespread view. Ran Bhagat of the International Institute for Population Sciences, across the city in Chembur, said, "The doctors are only interested in making money, and they can get a fee of forty thousand rupees for a sex test and abortion. In Mumbai even ordinary doctors are wealthy, with their air-conditioned nursing homes for the rich."

Sex-selective abortion is a crime that is committed an estimated one million times a year in India. Yet prosecutions are rare, and I could find only one doctor who has been imprisoned. In 2006 Dr. Anil Sabhani and his assistant both got two years following a prosecution brought in Faridabad, near Delhi, after telling a pregnant undercover investigator that her fetus was female and offering to "take care of it." The Sabhani case took three pregnant investigators, video footage, and a five-year court process. And they got a conviction only because the doctor was careless in verbally declaring the fetus was a girl. The court heard one satisfied customer with three sons say there was usually a code. "If the doctor tells us to come and get the [ultrasound] report on Monday, we know it's a boy. Friday means a girl." Other local doctors signed forms in red ink for a girl and blue for a boy. Usually, said the judge, "no words are exchanged. It's an unspoken thing, and one doesn't even have to ask."

It is controversial in India to suggest that the epidemic of sex-selective abortions is related to reduced family size. But it clearly is, says Sista. "When people had more children, they would simply keep trying till they had a boy. Now most couples are having only one or two children, and they want to be sure they have a boy." Indian community health expert Mohan Rao agrees: "Given the ideology of son preference, the government's vigorous pursuit of a two-child norm is an invitation to sex-selective abortion." Every reduction in fertility brings a reduction in the proportion of girls born. By one estimate, published in the *Lancet*, India has missed ten million girls in the last twenty years.

Reporting the Sabhani conviction, journalist Carla Power said, "India's female feticide problem is entwined with the consumer society. If one can order a BMW, goes the mindset, one can order a boy." Everyone I talked to had a nervous anecdote about women they knew who had had abortion after abortion until they finally got a boy. I was told, "If you go to Delhi, ask around the World Bank offices about how many of the staff there have daughters, and you will be very surprised. There are virtually none."

This is shocking news for the status of women in India. But Namita Sharma of the University of Jammu sees women not as victims but as perpetrators. "Educated mothers are more efficient at discriminating against daughters," she says. From Delhi in the north to Tamil Nadu in the south, the better educated women are, the fewer girls they have. In a study of middle-class couples in the northern state of Punjab, 73 percent of women said that if a sex test showed a fetus was a girl it should be aborted, compared with 60 percent of men.

Punjab is the heartland of sex-selective abortion. There is not a single district in the state where more than eighty-two girls are born for every hundred boys. The districts dominated by Jat Sikhs have the lowest ratios. Jats are the Sikh agricultural nobility. In Khamano, where Jats form two-thirds of the population in a neighborhood described by the *Sikh Times* as "rich in buffaloes and boys," there are just sixty-three girls for every hundred boys.

Neighboring Haryana state is little better. Locals say that back in the 1980s and 1990s, doctors toured the villages with mobile ultrasound units, advertising their services with the slogan "Pay 500 rupees now and save 50,000 rupees later." Fifty thousand rupees was the going rate for a dowry. The pitch proved so successful that in the 2001 census there were only eighty-two new girls in Haryana for every hundred new boys.

Everybody agrees more girls are needed to redress the imbalance, but few are willing to give birth to them. With girls in short supply, some families enter into closed-shop agreements in which they agree to marry off their sons and daughters to each other. But still, spare sons are usually left behind. Many of them head south to find girls in more egalitarian states like Kerala. Rambir, a thirty-three-year-old from Sorkhi village in Haryana, didn't want the bachelor fate of his four older brothers, so he asked a well-traveled neighbor to act as a go-between to find a bride two thousand miles away in Kannur, Kerala. As the *Times of India* described it, "Such nuptials are usually quick—with neither family understanding the language of the other. The groom's family pays for the train travel and the cost of the marriage itself, with another Rs 20,000 [around $400] thrown in to pay the go-between—usually a woman from Kunnar herself married to a Haraynvi." There are no big dowries in this bride market. The brides are usually poor but educated, for Kerala has universal schooling. Their husbands, by contrast, are ill-educated farmers with a bit of land.

Female feticide is the latest manifestation of a neglect of girls that spreads right across India. "If you lack resources, then what you have goes first to the boy child, whether it is food or money for health care or education," says Dr. A. L. Sharada, who works with Sista at Population First. "Even among the middle classes, girls are second-class citizens." Perianayagam Arokiasamy, from the International Institute for Population Sciences (who proudly told me he has one daughter and no sons), agrees: "It starts with nutrition and food. Folklore has it that girls need less, but boys have to grow up strong. So girls get sick more, and their diseases are ignored, or they get home remedies rather than modern medicines. Girls don't get as much immunization."

Sometimes neglect turns to something worse. Stories of baby girls left to die in the fields may make headlines now. But much of what goes on never makes news. In Haryana, young girls are more than twice as likely to die as young boys—a difference "higher than in any country in the world," says Namita Sharma. And the chances of making it to adulthood are lower the more older sisters a girl has. The desert state of Rajasthan is notorious for outright female infanticide. According to Sharma, it "continues even in the family of a prominent Rajasthan politician." She doesn't name the family, but says that in the late 1980s, no girls had been registered as born in the family for forty years.

The shortage of women in Rajasthan is reported to have triggered an epidemic of kidnappings of girls and the widespread sharing of wives. Punjab also reports cases of brothers buying a woman from a poor family so she can become wife to them all, and sometimes to their fathers as well. The practice is called *draupadis*, after a woman in the classic Indian epic *Mahabharata*, who married five brothers. Such arrangements are likely to become more widespread. With sex ratios still falling, the big boom in surplus males has yet to reach puberty. By 2020, there will be 25 million surplus Indian men, mostly looking for wives.

Former UN chief demographer Joseph Chamie predicts that "sex imbalances will push many men to look for child brides." He warns of bachelor gangs creating crime and disorder, trafficking women, and committing rape, and even "the build-up of large militias to provide a safety valve for the frustrations of numerous bachelors." There may be hysteria in this. But the danger is that economic progress could mean yet fewer girls. In India the phenomenon of missing girls is "not the outcome of poverty," says Sharma, "but of relative affluence; not the result of illiteracy and ignorance among women, but of education."

· · ·

Is India alone? Well, no. Indian-born mothers in Britain appear to have begun taking holidays in India to have sex-selective abortions. The number of girls born to this group has fallen from ninety-six for every hundred boys in the early 1990s to just 92.5 today, with most of the change in the sex ratio concentrated among second and subsequent children.

In 1992, shortly after the collapse of the Soviet Union, the sex ratio of babies in the newly independent Caucasus states of Armenia, Azerbaijan, and Georgia all began to drop. They reached between eighty and eighty-five girls per hundred boys at a time when fertility was falling fast. In Georgia, where fertility is now 1.4 children per woman, the sex ratio for a third child is only seventy girls for every hundred boys. In Armenia, where the fertility rate has been as low as 1.1, there are only around fifty girls for every hundred boys among third children. Not even India can match that startling figure.

But the most missing girls are in China, where only eighty-four girls are born for every hundred boys. The "main reason," the head of the National Population and Family Planning Commission, Zhang Weiqing, told journalists in 2005, "is the existence of thousands of years of a deep-rooted traditional view that men are worth more than women." Maybe, but its revival is a product of the one-child policy.

In Confucian societies, family rituals should be led by the eldest son. It's a bit more than carving the Thanksgiving turkey. If no sons are born, the family dies.

The poor Han people of southeast China and Taiwan, in particular, have a long and distressing tradition of female infanticide that skewed the sex ratio right up until the advent of communism. It had a name: *Sheng Zi Bu Ju*, meaning having children but not bringing them up. Drownings, suffocation, or simply leaving a wrapped bundle exposed on wasteland (things that in the West would bring prosecution if you did them to a cat) were endemic.

Chinese demographer Zhongwei Zhao says the historical literature is full of such reports. Infanticide of both boys and girls was often the major means of limiting family size—"regarded as unfortunate but understandable; sinful but not punished by law," says demographer Jack Caldwell. But girls were sacrificed the most. Han peasants had a chilling saying that demographers like to quote: "The birth of a boy is welcomed with shouts of joy and firecrackers; when a girl is born, the neighbors say nothing."

In the 1930s, China had about 8 percent more men than women. Mao's baby-boom policy from the 1950s brought the sex ratio among newborns back to normal. But in the early 1980s, female infanticide resumed with the one-child policy. "Highly abnormal sex ratios in Guangxi, Guangdong and Hainan in the south, Shanxi in the north, and Anhui and Zhejiang in the east . . . all pointed to extreme maltreatment of very young girls at that time," says New York–based China specialist Judith Banister.

Another American China watcher, William Lavely of the University of Washington, has written a searing account of events in Diandong county, in the remote southern province of Yunnan. County statistics show that between 1989 and 1991, infant mortality was between two and three times higher among girls than boys. The death certificates for boys record a specific illness, but most of those for girls simply say "accident" or "unknown."

Something horrible was clearly going on. One local health worker admitted to Lavely's researchers, "When a male is born, everything is fine, but when a female is born, they throw the baby away. After seven or eight days they report that the baby has died. There are a lot of these cases." Another said, "Some women just keep having children until they have a boy. The girls they just get rid of." Nobody would argue that Diandong was typical. But it may not be unique. In neighboring Guangxi Province, the mortality rate among young girls remains double that for boys. Lavely says his findings certainly contradict researchers who deny that infanticide occurs in modern China.

In most of China, sex-selective abortion following ultrasound scans has largely replaced female infanticide. In 2000 China nationally had only eighty-five girls born for every hundred boys. By 2006 the ratio had fallen again, to about eighty-three. Among second children, it was just sixty-five. But this is no direct substitution for feticide or infanticide. The geographical pattern is different. Sex-selective abortion is concentrated among the rich, industrialized, and urbanized provinces—where incomes have been rising fastest, and where the one-child policy has been most rigorously enforced. Hainan and Guangdong could manage only seventy-three and seventy-seven girls, respectively, for every hundred boys in 2000. The only Chinese provinces with normal sex ratios today are those dominated by minority groups, for whom the one-child policy does not apply.

China's "missing girl" count over the past thirty years is now estimated at 25–30 million. Like India, China is becoming a nation of surplus males. The "little emperors" are now turning into a bachelor's army. Unlike in India, however, sex-selective abortion is not a crime in China. This is controversial, but many think it is a good thing. If it were banned, says Lavely, infanticide would be sure to return. But meanwhile, pity the young men with no young women. Increasingly, says Zhongwei Zhao, the thin ranks of nubile young women will be snapped up by the country's older and richer men, leaving more and more younger prospective bridegrooms on the scrap heap before they begin. Lost little emperors.

· · ·

They know all about this in Korea, which has a similar Confucian tradition in which boys are strongly preferred because they are expected to stay in the family home with their aging parents. Girls head off to the in-laws, accompanied by a hefty dowry.

After the end of the Korean War in the 1950s and the division of the country into communist north and capitalist south, couples in South Korea began to reduce their family sizes—not out of compulsion but through rigorous persuasion, says Korean demographer Minja Kim Choe, who was nine years old when the country was divided. The fertility rate went from around 6 in the 1960s to 1.2 today. As it fell, ultrasound equipment began to be manufactured in South Korea, and couples started to take advantage of the new technology to check the sex of their fetuses and abort the girls. In this way, the traditional family's need for a boy was reconciled with population policies. But the problem of missing girls grew.

In 1987 the government outlawed sex-selective abortion, but the rules were largely ignored until recently. Now South Korea is emerging as the first country in modern times to shrug off the curse of aborting girls. In the past five years, it has had a modest baby boom—in girls. The number of girls born for every hundred boys has risen from eighty-six in the mid-1990s to ninety-three in 2006. Minja says that "the traditional family system is breaking down, under the effect of economic change, increased higher education, and altered expectations, especially for women." Women have jobs of their own and little interest in marrying men who offer them a lifetime with the in-laws. As the old family system disintegrates, so do the obligations that came with it. In a recent survey, only a tenth of South Korean women felt any pressure to have a son to maintain the family line.

The recovery of the sex ratio among the newly born will take many decades to work its way through the Korean population, however. Surplus boys will be reaching adulthood for another decade or more. By 2020, according to Minja, there will be 123 South Korean men in their late twenties chasing every hundred women. Many South Korean men have long since found themselves surplus to female needs and have gone hunting for women across East Asia. Many end up in Vietnam, where the long Indochina wars left a dearth of men. Cross-border bride hunting is destined to be a booming business in Asia for years to come (see chapter 20).

The South Korean experience offers hope that other countries in Asia may also eventually turn the corner on sex-selective abortions. Like ultralow fertility, female feticide looks like the result of an incomplete sexual revolution, in which sons are still revered and daughters downgraded, even by their mothers. For now, widespread female feticide across Asia offers an alarming refutation of some easy Western assumptions about development and modernization.

CHAPTER 17

Where Men Still Rule

I put on a skull cap before I went in. "You have to go incognito," said my guide, Rabbi Carmi Wisemon. The college was bursting with men shouting loudly in Yiddish and Hebrew. The noise filled the corridors and cloakrooms and echoed down the stairs, through the libraries and vestibules. In one lecture theater, almost a thousand students were bellowing in pairs at each other as they read aloud and debated the meaning of ancient religious texts, known as the Torah, as if they were hot off the presses. Behind them, the walls were lined with the sacred books, all heavily rebound and repaired after constant use.

Many of the seven thousand students in Mirrer Yeshiva, and the tens of thousands more in other theological colleges close by, have been doing this every day for twenty years. They have come from all over the world. I met students from Brazil and New York, Manchester and Odessa. They are the ultimate permanent students. Except that, in their black, wide-brimmed hats, waistcoats, jackets, white shirts, thick beards, and side curls, they don't look like regular students.

I was in the neighborhood of Mea Shearim, the heart of world Jewry, just a block from Jaffa Street in central Jerusalem. This is the home of the ultraorthodox Haredi, who have created a "society of scholars," in which around two-thirds of Haredi men between ages twenty and fifty-five study rather than work, including almost all males under forty. They are largely dependent for cash on charity and welfare payments from an Israeli state about which they are so ambivalent that they refuse to do national service.

Mea Shearim is not just the praying heart of Jewry; it is also its breeding heart. For those male students, despite their segregated studying, are all married. Outside, their wives, white-faced women in head scarves and thick stockings, were herding large broods of children through the streets. Families of eight or more are quite normal here.

The streets were packed. Eight-year-olds pushed two-year-olds in battered buggies, steering into and out of the tiny stores containing basic kosher provisions and a little cheap junk food, then careering past secondhand furniture stores and bookstores, brushing against posters pasted to walls declaring everything from the whereabouts of cheap buses to death notices and missives from rabbis. Above, on tiny balconies, were more gangs of children, minded by older sisters already looking like their mothers and playing amid huge lines of washing.

The word "ghetto" was originally coined to describe cramped, overcrowded Jewish neighborhoods in European cities. And now I know why. Many of these huge families live in two-room apartments. Mea Shearim is one of world's last bastions against the global trend toward fewer children. It is like going back in time, not simply because of the dress, but because few other city neighborhoods teem with children like this anymore. Cities generally have the lowest fertility rates in any country, and yet here the urban Haredi have among the highest rates in the world.

The Haredi men guard their enclosed world with zealotry. Big signs at either end of Mea Shearim Street tell visitors how to dress and instruct tour groups to keep away. Any drivers foolish enough to venture down here on the Sabbath are likely to be pelted with stones or—a new favorite—used diapers. There are plenty of used diapers.

This community is new, but dressed up as old. It has built up from just a few dozen pioneers in the 1950s to an estimated quarter million today. The Haredi make up a third of the population in Jerusalem and more than half of its young population. In total, Israel now has 750,000 Haredi, 15 percent of the Jewish population. And far from reducing their fertility like their fellow Jews, they are having more babies than ever before.

Spilling semen is a sin, say the rabbis, who rule the roost—so no condoms. And not much contraception of any sort, since another commandment is to "be fruitful and multiply." Parents pair up their youngsters before they are twenty. And once married, the women are expected to begin producing children straightaway. And to carry on. If pregnancy does not follow intercourse, couples quickly resort to in vitro fertilization treatment. Israel has the largest number of fertility clinics for its population size in the world, three times as many as the United States.

Environmentalists will be appalled at this "irresponsible" fecundity. Yet the Haredi have an environmental wing. Akiva Wolff, a soft-spoken American by birth, is a leading Haredi scholar, writing a thesis analyzing the environmental implications of a Torah commandment not to "needlessly destroy." He grew up in the United States as a nonreligious Jew. "It was the time of Paul Ehrlich and the population bomb," he remembers. Yes, living in a community with such dramatic fertility is "very different," he concedes. "I can see both viewpoints. From a Haredi perspective, children are the greatest blessing. We are not used to thinking about population control." But at least, he said, the Haredi have smaller ecological footprints than most secular Israelis. "It's not just because they are poor. It's a question of values. They want to pursue spiritual goals. Not many have cars. They are not consumers."

Why such a devotion to procreation? It is partly the trauma of the holocaust, Wolff says. Roughly six million of the world's Jewish population of seventeen million died in the Nazi gas chambers. Orthodox Jews were the greatest victims of the Holocaust roundups of Jews. The secular Jews had often got out, but the orthodox people listened more to the rabbis, who told them to stay. Ever since, Wisemon says, "there has been a sense of wanting to replace those lost. I am named after five of my grandfather's sisters, who were all murdered. In a sense, I replace them." Of course, "when they go to bed at night, they aren't thinking about history, but there is an underlying consciousness about it," agrees Wolff. "They had to rebuild, to make up numbers."

And rebuild they have. Prewar orthodox Jews typically had four children. Starting in the 1950s and 1960s, the orthodox Jews of the world began to congregate in Mea Shearim. They began to marry earlier and to have many more children than before—just as the rest of the world began its long journey to later marriage and smaller families. An average of five children in the 1950s became six in the early 1980s, eight by the mid-1990s, and nine by the turn of the millennium. The Haredi population doubles every fifteen years. They are not so much holding out against modern fertility norms as rowing hard in the other direction.

In 2006 a community Web site reported the death, at the age of forty-four, of Ahuva Klachkin, a woman who had given birth to eighteen children. Under the headline "How many children does it take to be righteous?" the article reported the emergence of "hundreds" of very large families that have "long since passed the 12-child mark." Women literally have no other role in society than producing and raising children. A local midwife was quoted as saying, "You can see [the mothers] on Geula Street. Most of them walk very slowly, and look very old, but they are actually fairly young."

Secular Jews are confused and sometimes angered by the Haredi. "I think they hoped that the Haredi would shrivel away. It has come as a shock that they are growing," says Wolff. Many regard the society of male scholars as scroungers, contributing little to the national economy but taking a lot in welfare payments for their children, and even avoiding national service by continuing with their studies.

But the fertility of the Haredi is changing the politics of Israel. The Haredi are not Zionists (one reason they avoid the draft), but their many foreign supporters see them as Israel's saviors. And one prominent American propagandist, Yoram Ettinger, thinks that their burgeoning numbers can hold the line in a demographic war with the Palestinians. Ettinger claims that they will allow Jews to remain forever the largest population between the Mediterranean and the River Jordan. "The current fertility levels of the Jewish population now constitute a strategic asset," he says. There need be no two-state deal with the Palestinians. Israel can use sheer force of numbers to maintain its hold on the entire territory.

Ettinger has been engaged in a long and acrimonious dispute on the matter with Sergio DellaPergola, Israel's top demographer. DellaPergola says Ettinger's argument is politically motivated nonsense. "There is only a small Jewish majority east of the River Jordan, and it will disappear." On average, despite the best efforts of the Haredi, Jews in Israel have 2.8 children per woman, compared with around 3.6 for Arab women in Israel and on the West Bank and 5.6 in Gaza. Recent census data suggest the Palestinian population rose 30 percent between 1997 and 2007, compared to an Israeli increase of 5 percent. On top of that, the Arabs are younger than the Jews, he says. "Therefore they die less. Ettinger ignores that. The Arabs have a higher fertility and a lower death rate. Put those two facts together, and their rising share of the population is certain."

DellaPergola's data were critical in Israel's decision to pull its settlers out of Gaza in 2005. Afterward, there was a furor in the Knesset when Ettinger claimed Palestinians were making up their fertility data to hoodwink the Israelis. Demography feels more immediately like destiny here than in most places.

Most of the more thoughtful members of the Haredi community see its ultrahigh fertility as a passing phase. "Nobody expects it to continue forever," says Wolff. "In the future the men will go back to work and the women will have fewer children." Menachem Freidman of Bar-Ilan University in Tel Aviv, who has studied the Haredi for half a century, agrees. "Frankly, I am surprised it has lasted as long as it has. Every society allows some people to take a spiritual path. But not so many." And usually those spiritual people don't have children, he points out.

The problem now is that by having so many children, the Haredi are gaining a lock on the Israeli political process. Their multitude of small parties are often seen as a joke by outsiders, but they have become essential members in a succession of coalition governments. And their bottom line is continued economic support for their society of scholars and its ever-growing band of members. "One day the government has to decide not to support the Haredi," says Friedman. Maybe, but logic doesn't often play a role in politics between the Mediterranean and the River Jordan.

·   ·   ·

Less than an hour's drive from Jerusalem is another male-dominated community with ultrahigh fertility: the Bedouin of the Negev Desert. Once a nomadic people, most Bedouin now live in seven "recognized villages" allocated to them by the Israeli authorities in the 1970s, or in forty-five "unrecognized villages" scattered in the desert with few services and little infrastructure. Wadi Nam is one of the largest unrecognized villages, a collection of corrugated iron shacks nestled under power lines between an industrial complex, a chemical waste dump, a power station, and the road from Beer Sheva to Eilat. Over the hill is the

Dimona nuclear plant. Some five thousand people live in Wadi Nam, though at the time of writing there were plans to move them on to make room for a large Israeli military camp.

When the Israeli state was established in the late 1940s, there were perhaps ninety thousand Bedouin in the Negev. Most fled to Egypt, Jordan, or Gaza, leaving behind about eleven thousand. But that small group has now grown to around 200,000, thanks to one of the world's highest fertility rates. While Palestinians on the West Bank typically have 3.6 children per woman, and even the beleaguered people of Gaza make do with 5.6, the Bedouin women in the Negev are up there with their Haredi sisters at 7.6 children each.

The root of this fecundity lies in the destruction of their way of life, says Amal El-Sana Alh'jooj, who was brought up in the Bedouin recognized village of Laqiya. When the Bedouin were forced into the villages, they lost most of their pastures and sheep. The men had to find work. Being unskilled at anything but herding, they were bottom of the pile and joined work gangs. But Bedouin women lost out even more. When they were nomads, Bedouin women "were partners in the desert," says El-Sana Alh'jooj. "There may not have been equality between men and women, but the women couldn't be ignored." Now, she says, they are "imprisoned within four walls" in masonry homes in the townships. El-Sana Alh'jooj, one of thirteen children, has seen her mother's life change for the worse. "Once she helped look after the sheep and would greet guests who came to the tent, because it was her domain. But in her house she doesn't feel able to do that without her husband being present. Her sense of equality has gone backwards."

For many women, their role is now limited to childbearing, she says. "They have nothing else to do. My mother says that giving birth is the only way Bedouin women have left to express themselves. And it is the only way to win respect from their husbands." Some of the highest fertility rates are in unrecognized villages where there are no high schools, so "girls have little else to do when they grow up other than have children." Many girls still marry at fourteen or fifteen.

El-Sana Alh'jooj left her village and is now director of an NGO called the Arab-Jewish Center for Equality, Empowerment and Cooperation. She is one of the leading figures in the fractured tribal Bedouin community. Another is academic Alean Al-Krenawi, who is professor of social work at Ben Gurion University in Beer Sheva, the Negev's largest town. "The Israeli government never prepared the Bedouin for the modern life it imposed on them," he says. "It's hit everybody hard. Women have lost power. They are the poorest of the poor in a supposedly Western country."

On the face of it, there are fewer reasons today for the Bedouin to have lots of children. There are no sheep for the youngsters to look after, so they have no economic role. Nor is there a need for lots of boys to grow up and fight the many skirmishes that once took place between tribes over land. But Al-Krenawi, who has only two children himself, says that from a male perspective, a large family is still a good way to political power. "I come from a big family in Rahat. Because of that, my brother became mayor." Unconsciously, he thinks, the powerlessness of the Bedouin today "encourages people to have more kids."

Things are changing. In some Bedouin villages, there is an explosion of interest in educating girls. Some 15 percent now attend higher education. There are more Bedouin girls than boys on the Ben Gurion University campus. This changes their outlook, and some tribes are cutting their fertility fast as a result, says El-Sana Alh'jooj. "My tribe, the El Sana, has seen a huge change in attitudes. I have twins, and my sister has three kids, rather than thirteen. She says kids are not bringing her anything. She wants another life." Older women, too, are looking for ways to become more independent. I visited a wool-weaving project in Lakiya village, which employs 150 women from several villages making rugs and bags.

But it's not just women who are asserting new rights. Men are too, with unpredictable results. One is polygamy, which has lately become rampant. "In the past, only the head of the tribe had many wives. That's how it was in my father's generation," says El-Sana Alh'jooj. "Now thirty percent of families are polygamous. Boys often have to marry their cousin when they are young. That's the tradition. But when they go outside the tribe to find work, they see other women—often Palestinian women from the West Bank and Jordan. Then they say they want a second wife they have chosen for themselves. They see it as their right." Polygamy is, of course, another reason for the soaring Bedouin population.

Two-thirds of the Bedouin in Israel are under twenty years old. More than a fifth are under five years old. This makes the Bedouin one of the youngest populations in the world. Unemployment is already very high, at 50 percent among men and 90 percent among women. "If we don't invest in educating and taking care of the younger generation, there will be more crime and drugs. We are already starting to see it," says Al-Krenawi. He predicts there will be "an intifada [uprising] in the Negev" among the Bedouin youths. "They will think they have nothing to lose and will start to destroy." El-Sana Alh'jooj agrees. "The Israelis see us as a demographic threat, as well as a security threat. But this is a self-fulfilling prophesy. It will happen if they continue to blame the victims of what is going on here. We have to give our youth a stake in the future."

The Haredi and Bedouin seem, on the face of it, entirely different communities. One urban and one rural. One choosing to be a "society of scholars," the other stuck between traditional nomadic ways and a new settled life they did not choose. One with political power; the other with effectively none. But both are victims of profound social trauma that has left them fighting to retain a sense of identity. Both, consciously or not, have taken refuge in numbers, finding importance and meaning in ever more people. Both, above all, are characterized by male rule over reproduction.

PART FIVE

·

MIGRANTS

In the midst of the peoplequake, people are on the move as never before. As workforces in Europe and parts of Asia implode, surplus people in struggling economies are eager to fill the gaps. And travel has never been so easy. If you have a thousand dollars and a tourist visa, you can leap on a plane, be almost anywhere in the world within twenty-four hours, and disappear into the throng. The migrants are badly needed. And they are the advanced guard of an increasingly globalized workforce of transnational citizens. But even as they become essential to the economies they join, they are demonized and criminalized, making them vulnerable to criminal gangs and worse.

CHAPTER 18

Waving or Drowning?

"Sinking water, sinking water. Many, many sinking water." The voice of Guo Binlong, screaming into his mobile phone through the sound of wind and rain, summed up the desperate life of one Chinese migrant. And it seemed to speak for millions more. As the freezing waters of Morecambe Bay in northwest England rose up past his chest, Guo began gulping water. The police phone operator recorded it all. But the man's faltering English had no way of explaining where he, or his forty fellow cockle pickers, were. They had been harvesting the clamlike shellfish in England's most notorious estuary, where tides come in faster than a galloping horse. The cockle banks and most of the cockle pickers were engulfed in seconds.

Before he died, Guo also phoned his mother, five thousand miles away in Fujian Province in eastern China. He asked her to pray for him. "The bosses got the time wrong. I can't get back in time," he told her. By the time she had rushed down the lane to tell his young wife, Yu Lihong, of her husband's peril, Guo had disappeared beneath the freezing waters.

It was the night of February 5, 2004. In Fujian, several of Yu's neighbors soon learned the terrible fate of their own husbands on the English cockle beds. The Lancashire police's recording of Guo's final plea was heard at Preston Crown Court two years later. Lin Liangren, the Fujianese subcontractor, or "gangmaster," who employed the cockle pickers and sat in his car on the beach as they drowned, was convicted on twenty-one counts of manslaughter and of trying to persuade survivors to deny that he was their boss. He received fourteen years in prison. It emerged that while he paid the pickers eight dollars for every bag of cockles, he sold them to local traders for twenty-four dollars. "All I wanted was quick cash," he told the court.

That was why Guo was in England too. Thin and gangly, Guo was one of life's losers. An eldest son in a country that now rarely allows couples more than one son, he saw his younger brother make money by traveling to Japan to work. Like tens of thousands of other fortune hunters from Fujian, his brother had gained the trappings of wealth and built an ostentatious house for his family. Guo had stayed at home with his wife and started a brick factory. It failed. He had opened a gas station. That too went belly up. He borrowed money from his father, and borrowed again. He was in his late twenties, with a four-year-old son, and had lost face badly. So he decided on one last throw and went to see the people smugglers known as snakeheads.

By the time Guo reached London, his family back home had paid out some thirty thousand dollars in installments to the snakeheads. But he failed to find a job in a food-packing factory or a Chinese takeout restaurant or any of the other places where fellow Chinese migrants to Britain usually find work. In the end, he agreed to go to Liverpool, where he found himself sleeping in shifts with thirty others on the floor of a flat above a shop in the tatty inner-city district of Kensington. Several times a week, gangmaster Lin drove them in a white minibus through the night to work on the cockle beds of Morecambe Bay. They are reputedly among the finest cockle beds in the world, but in one of the most dangerous bays, with ripping tides, quicksand, and fast-flowing rivers coming down from the Lake District.

The work was difficult and dangerous. But there was a Klondike fever that winter. The bay had the biggest cockle harvest anyone could remember. The English cockle traders were as ruthless in their demands, and as reckless in their disregard for safety, as the gangmasters. Some bought cockles directly at the beach. Word spread, and large refrigerated trucks arrived from Spain. Out on the banks, hundreds of cockle pickers from both local and Chinese gangs jockeyed for territory. According to one local trader, a group of Lancashire fishermen had set fire to piles of cockles collected by Lin's gang.

All the elements for catastrophe came together on the night of the tragedy. Low tide, when the cockle banks were exposed, was at night. It was raining, so the outflows of the rivers were running strongly across the bay. The wind picked up, so the incoming tide arrived even faster than usual. And just before the tide turned, Lin's gang found rich pickings miles offshore. Other teams thought it was too dangerous to work and had left. But for Lin, safe on shore and giving instructions by mobile phone, this was an opportunity to avoid the local gangs who had been harassing his team.

But Lin kept the gang out too long. The bank where they were working was surrounded by cold, raging waters. Before they knew it, the route back to shore was flooded. Guo Binlong, Yu Hui, Xie Xiaowen, Guo Nianzhur, and many more all died. In the morning, rescue teams found their bodies huddled together on the bank. Back home in Fujian, families were left to grieve, receive the bodies, and try to pay off the debts. Guo's mother committed suicide soon after.

Fujian Province is probably the people-migration capital of the planet. Around a million Chinese have taken one-way flights from the airport at Fuzhou, the provincial capital, since the late 1980s. They head for Chinatowns in New York, São Paulo, Belgrade, Dubai, Johannesburg, or most often, in the capitals of western Europe. It is an old trade, dating back at least to the nineteenth century. The snakeheads that organize their often complicated journeys operate with the tacit support of Chinese local authorities, who foresee remittances boosting the local economy.

Snakeheads are essentially travel agents whose extras command a high premium. They are named after an Asian fish that can slither across dry land from one pond to another—a fair analogy for what the people traders do for their clients. They may often operate illegally, and they subcontract some legs of the journey to criminal gangs. But they are smugglers of willing people, not brutal traffickers of the duped, drugged, and terrorized.

Frank Pieke, director of Chinese studies at Oxford University, has studied the Fujian people trade. He says local teachers run government-sponsored training classes that amount to bluffer's guides on how to get by in the foreign lands to which the snakeheads make deliveries. Enough villagers have returned that would-be migrants know the conditions they can expect in transit and on arrival in Europe. The families pay a fifth of the cash up front and the remainder on delivery. Typically they borrow the money, and the migrants send remittances back home to pay off the loan.

Mostly it works. Otherwise people wouldn't go. In the paddy fields of Fuqing and Chandle, you can see hundreds of five- and six-story neoclassical houses, built with the cash earned after the snakeheads and moneylenders have been paid. In some villages, most families have someone working abroad. Even menial jobs like cleaning dishes in a Chinese restaurant can build a big house back home. Most of the time, neither the migrants nor their families would describe the snakeheads as "trading in human suffering," the common currency of European newspaper articles on their activities.

But things can go badly wrong. At least sixteen of the dead cockle pickers came from central Fujian. This was also the homeland of another fifty-eight Chinese whose deaths made headlines in Britain in 2000. The group flew into Europe via Slobodan Milosevic's Belgrade before being smuggled to Zeebrugge in Belgium and loaded into a container on a truck, where they were found suffocated on arrival in Dover.

In Britain, many Fujianese work for Chinese restaurants, though they rarely get to own them, thanks to the determination of the Cantonese from Hong Kong to maintain their century-long control of the business. But more and more Chinese in Britain are failing to get even the modest jobs that once justified the snakehead fees. Many end up picking vegetables or selling DVDs on the streets of small towns. And there are some three thousand migrant Chinese women who, according to author and *Guardian* writer Hsiao-Hung Pai, work in the sex industry in England. There are, she says, six hundred Chinese "massage parlors" in London—half the number of Chinese restaurants.

· · ·

Would-be migrants fly if they can afford the fare. Those who can't afford to fly take a boat. The term "boat people" became famous in the late 1970s, when tens of thousands of Vietnamese fled their homes for the then-British colony of Hong Kong. The phrase is now used round the world for people fleeing poverty or persecution, seeking wealth and freedom. For five decades, Cubans have been sailing small boats across the Florida Strait to the United States. Southern Florida is now full of them. Likewise, the Caribbean island of Haiti sees periodic exoduses.

The tragedies are frequent. Hundreds die every year when storms hit dhows carrying Somalis and Ethiopians toward the construction sites of the Persian Gulf. Often, after the fish have had their fill, they leave behind little more than a pair of sneakers or a belt. In 2001 there was international outrage when 353 asylum seekers sailing from Indonesia to Australia drowned when their vessel sank, despite daily air monitoring of their passage.

The past few years have seen thousands of stateless Rohingyas people drowned in the Bay of Bengal as they escape the Burmese generals. Most have been heading for Thailand or Malaysia. In January 2009 hundreds of them were detained by Thai soldiers on Koh Sai Daeng Island; they were eventually put back out to sea in boats without engines, water, or food. Some five hundred were rescued after reaching India's Andaman Islands, but survivors said many more had perished in what appeared to be tantamount to murder by the Thai military.

But the steadiest and deadliest flow of boat people is across the Mediterranean from North Africa to Europe. On Christmas Day 1996, 280 migrants from Tordher, a Pashtun village in Pakistan's North-West Frontier Province region, drowned in the Mediterranean, apparently after being moved at sea from a Honduran-registered freighter to a Greek boat. They were en route to Italy, where a village elder had settled twenty years before and encouraged others to follow.

These tales of death on the high seas rarely make the news. They are routine. In August 2008 the drowning of seventy migrant Sudanese and Eritreans in a dinghy caught up in rough seas between Libya and Malta caused barely a ripple, even in the middle of the summer holiday season. After all, it came only a day after thirty others died in a boat off Spain. Boat people? They get what's coming to them, some say. But the tragedy is that Europe needs people, and the boat people's fates reflect the continent's hypocrisy.

For most of the past millennium, Europe has exported its people and its way of life to the world. America is largely composed of migrants from Europe and, via European slave ships, from Africa. But Europe now needs to import people. As its indigenous labor force shrinks, the continent is becoming a magnet for people seeking work. They pick oranges in Spain, olives in Italy, and cabbages on the English Fens; they serve drinks in London bars and Torino cafés, and open their legs in brothels from Munich to Manchester to Moscow.

Moroccans began to take the short boat trip across the Straits of Gibraltar to Spain in the 1960s. They were soon joined by Tunisians, Algerians, and Libyans taking boats to Malta, Italy, Cyprus, and Greek islands. By the 1990s Africans from Nigeria, Ghana, Senegal, and elsewhere in West Africa began to arrive in the Arab countries bordering the Mediterranean to take jobs in construction, in farming, or on oil fields. Some came by camel trains operated by desert nomads, others in Toyota pickups. Their guides were Tuareg nomads who took them along traditional routes through the desert cities of Agadez in Niger and Tamanrasset in Algeria, or via Sebha, the old Italian desert fort town in southern Libya. "Trans-Sahara migration has helped revitalize desert towns," says Hein de Haas of the International Migration Institute in Oxford, who has studied the migration. In Libya today, a quarter of the country's jobs are done by migrants.

After about 2000, the sub-Saharan Africans began piling into the boats to Europe too. They now take most of the seats. Even Chinese and Indians have flown to Accra or Bamako in order to join them. The number of migrants making landfall on Italian beaches rose 50 percent in 2008 to around 35,000, with at least another five hundred known to have drowned.

Europe has tried to keep them out. But as the patrols got tougher, new sea routes have opened up, from the western Sahara, Mauritania, and Senegal out into the Atlantic to the Spanish-owned Canary Islands. Local fishermen, who had lost out to European trawler fleets moving in on their rich West African fisheries, changed trades and began selling space on their wooden boats to deliver migrants to the beaches of Lanzarote or Tenerife. The journey can take two weeks, but once there, you are in the European Union. In 2007, 31,000 Africans tried this route. According to UN estimates, some six thousand didn't make it.

Meanwhile, Palestinians and Afghans, Iranians and Lebanese have begun traveling overland to Turkey and taking a forty-minute boat ride to Greek islands like Lesbos and Samos, where the Greek authorities have set up a detention center. Like their African counterparts, these migrants make asylum requests and then hope to disappear into the night.

Most migrants to Europe, whatever their legal status, find employment, because they are needed. Illegality often suits employers, who can treat people with no legal rights in virtually any way they want. But these "invasions" also allow politicians and much of the media to engage in reckless rhetoric against them.

The xenophobia is especially bad in Italy, where migrants are essential to a range of industries, including bringing in the harvests. In 2008 the mayor of Rome promised to deport twenty thousand immigrants whom he said had criminal records. North of Naples, a summer of growing tensions between locals and African migrants culminated in sixteen Africans being murdered in two separate attacks on their camps, allegedly by the Neapolitan mafia, the Camorra. In addition, a poll found that two-thirds of Italians wanted all the country's 150,000 Roma expelled, even though most are Italian citizens. Prime Minister Silvio Berlusconi's promise to uphold "the right of Italians not to be afraid" sounded like carte blanche for vigilante squads to take up arms against the Roma. He did not mention the rights of Roma not to be afraid.

The frequent deaths among migrants crossing the Mediterranean to work in Europe—like the deaths of the Chinese cockle pickers in Morecambe Bay—are a scandal that should embarrass the continent, says de Haas. They arise from the failure of Europeans to recognize their reliance on the people who want to join them. There are today almost 3 million Moroccans, 1.2 million Algerians, and 700,000 Tunisians living in Europe, along with numerous other groups. Most have jobs. Europe needs them. Ever more of them. So they come, only to be treated as invaders.

Most migration from Africa to Europe is "demand driven," says de Haas. The bush telegraph across Africa is sophisticated. Far from having romantic ideas about Europe, most migrants have a pretty clear grasp of what they face and why they are needed. But while some obtain citizenship, an increasing number exist in a legal limbo created by Europe's hypocrisy. It is absurd to make migrants take leaking boats across the Mediterranean when, if they did not come illegally, Europe would have to send for them.

The idea of a fortress Europe under siege from would-be migrants may be convenient for the many politicians who do not want to discuss the economic and demographic realities. It may be worse. "The current practice of barring most legal migrants while letting millions of others slip illegally through the back door serves only the worst interests and promotes lawlessness," says Gregory Maniatis of the Migration Policy Institute in Washington, D.C. "Smugglers are enriched, immigrants work in oppressive conditions, and xenophobia runs rampant."

Europe is far from alone in its fear of the foreigners that it needs to maintain its economy. In the 1990s Australia's One Nation Party, headed by right-wing Queensland demagogue Pauline Hanson, had brief electoral success with its demand for an end to immigration. And environmentalists such as Tim Flannery, a former Australian of the Year, have claimed that the country is overpopulated. Yet Peter McDonald, the Australian government's top adviser of demographic issues, says hostility to Asian migrants is potentially damaging to Australia. "In the next twenty years, immigration is the only means available to meet aggregate labor demand in Australia," he says. As the country's baby boomers retire, only migrants can maintain the country's economy. In a report presented to his country's immigration department in March 2009, McDonald warned that shortsighted and short-term decision making was bad for the country as well as for migrants.

Similarly, the World Bank estimates that Europe's labor force will decline by at least seventy million by 2050, while the labor force in the Middle East and North Africa will increase by at least forty million. Europe will need many of those people. The new relationship has not got off to a good start.

CHAPTER 19

Migrant Myths

As a citizen of London, I live in probably the most international city on earth. Walking down almost any street, I hear a babble of different languages: Portuguese and Swahili, Japanese and French, Swedish and Bantu, and any number of versions of pidgin English. Less than a mile from my house is the largely Muslim enclave of Tooting. A mile in the other direction is a colony of young, mostly white South Africans in Earlsfield. Equally close, there are Poles in Balham, Jamaicans in Battersea, and Filipinos in Hammersmith. All across London you come across unexpected ethnic communities: Koreans in New Malden, Somalis in Wembley, Ghanaians in Tottenham, Turks in Green Lanes, Portuguese in Stockwell, Sierra Leoneans in Southwark, Vietnamese in Hackney, Australians in Earls Court, and Bangladeshis in Brick Lane. It is a kaleidoscope that may fill some with dread, but which fills me with hope.

The people of the world are on the move—whether it is Mexicans crossing the Rio Grande into the United States, African boat people riding the Mediterranean swell into Europe, maids from the Philippines negotiating paperwork in Singapore, Bangladeshis flying to construction sites in Dubai, Chinese forestry workers slipping into the Russian far east, Ethiopian Jews heading for Israel, ethnic Russians returning from old Soviet states of central Asia, Indonesian nurses running English care homes, Zimbabweans fleeing to South Africa, German sex tourists settling in Cambodia, Somalis in Kenyan tented antechambers waiting for a new life in the United States, Malawian babies adopted by wealthy Europeans, Asian sailors touring the fleshpots of Rotterdam, or bounty hunters from the Dominican Republic washing up in Panama.

The great majority of the world's people still live and die within a few miles of where they are born and never have a passport. Most people who uproot move within their own country—mostly to the big city. China's army of migrant workers—one in ten of the country's population—in its new industrial cities represents probably the planet's largest ever peaceful migration. But an estimated 200 million people are currently living outside their country of birth—twice as many as thirty years ago.

Most international migrants join people they know or who come from their locality. Most are under thirty years old when they move, and they are at least as likely to be female as male. They are not generally the poorest in their home countries. One-third have the equivalent of a college education. But many will start on the bottom rung in their new domiciles, doing jobs known sometimes as 3D jobs—dirty, difficult, and dangerous—and sometimes as 5D jobs, with domestic and dull added.

Often the journey is a flip across the border. The biggest flows of migrants remain overland: within the European Union, from Mexico and Canada to the United States, and across the porous borders of Africa. Others may cross the world, following familiar, often old colonial, routes. Algerians still go to France, Kenyans to the UK, and the people of Suriname to the Netherlands. Congolese jump on flights from Kinshasa to Brussels. Spanish word of mouth can still alert Ecuadorians that there are jobs in Spanish orchards, and Brazilians still end up in Portugal. Some Punjabi villages in South Asia have emptied as street after street departs for Britain. The majority of workers in Britain's eight thousand–plus "Indian" restaurants come from Jagannathpur in the Sylhet division of northeast Bangladesh.

Some places are traditional exporters of people. There are eight million Filipinos working abroad at any one time, mostly men crewing merchant ships and women in nursing and domestic service. There are also eight million Mexicans north of the Rio Grande. A sixth of the work force of India's richest state, Kerala, sleeps in a foreign land—mostly amid the construction camps of the Persian Gulf. On a typical day, the state's main airport at Kochi sees off twenty-one flights to the Gulf, but only two to other foreign regions. There are also new sources of migrants. Tajiks were once stay-at-homes in the mountains on the old Soviet border with Afghanistan and China. But in 2007 one in five Tajik workers were abroad, mostly in Russia.

Some migrants are fleeing local overpopulation or environmental degradation (so-called environmental refugees). Sometimes there is a mix of push and pull. But most are simply looking to better themselves and their families' futures. They are economic migrants.

Among traditional importers, the United States has 38 million recent migrants, or 13 percent of its population, and probably many more illegals. Singapore would be 40 percent smaller without migrants, and Australia and New Zealand 20 percent smaller. Saudi Arabia and the other Gulf States between them currently employ 13 million foreigners, mostly from South and East Asia. Four-fifths of the residents of Qatar were born somewhere else. As a result, adult men there outnumber adult women almost three to one.

The whole of the Middle East is in a ferment of migration. Egypt has some 3 million refugees from civil wars in Somalia. Palestinian and Iraqi refugees make up a third of the population of Jordan. Syria's diaspora is estimated at 20 million and Lebanon's at 15 million—more in each case than are left at home. Israel, of course, is a country of immigrants, and it has created many refugees as its people have annexed land across the former Palestine. Having penned Palestinians into Gaza and the West Bank, Israel now employs Romanian construction workers, Thai farm laborers, and Filipino nurses.

Many migrant journeys are much less permanent than they once were. In the "golden age" of nineteenth-century migration, the trip in a ship across the ocean was an arduous and probably a once-in-a-lifetime journey. But in the twenty-first-century global village, those with the papers often go back and forth almost with the ease of civil servants taking the train from Weybridge or PR executives coming in from Westchester. We are seeing the rise of the transnational citizen, without exclusive loyalty to any country. They live in a world of constant text messaging, e-mailing, and Skype calls, of electronic transfers of cash and frequent flights home.

. . .

That, I admit, is a somewhat idealized version of world migration. Here is some of the rest. A significant minority of migrants are trafficked against their will, through force, blackmail, or gross deception. Most victims are young women, forced into bar work, prostitution, or domestic service. In Saudi Arabia, household servants are sometimes treated "almost as slaves," according to Human Rights Watch.

The major trafficking routes are out of southeast Asia and from Eastern Bloc countries. In *McMafia,* his book on international crime in a globalized economy, Misha Glenny tracks some geographically bizarre people trades. They include young women from Transnistria, a self-declared independent enclave of Moldova, enticed to Egypt—via Odessa in Ukraine and Moscow—before being smuggled across the border into Israel to work in the brothels of Tel Aviv. He found Albanian, Macedonian, and Bulgarian women shoveled up a Czech "road of shame" to the Germany border, where Bavarians arrived with their euros to buy sex in their cars. In the bars of Dubai, established Russian prostitutes battled Chinese new arrivals. Glenny discovered that Moscow has become a major destination for women from West Africa, but that the former Soviet Union is also a source of migrants, with Azerbaijani women showing up in brothels in New Delhi.

Women are widely trafficked within Asia. Infamously, Thai and Cambodian mothers sell their daughters to urban brothel keepers for the delectation of Westerners. But there is also an international trade in Vietnamese women and girls sold to China, where they are put to work as prostitutes or become wives to grown-up "little emperors." They are a lot cheaper than Chinese brides, and Glenny notes that "for some poor Chinese families, it is the only way to find a spouse for their son."

The distinctions are blurred between slave labor, debt bondage, and a worker, whether man or woman, who has simply borrowed money in order to be transported to take up employment. Children, whose labor is sometimes given as payment for a loan made to the parents, are especially vulnerable. West Africa's child "chocolate slaves," sent from Niger to work in the cocoa plantations of the Ivory Coast, became notorious in 2001 after a shipload was impounded in Benin.

What begins as a contract freely entered into can become effective slavery if the debt cannot be paid off. In 2009 the Brazilian government announced that it had freed more than four thousand "slaves" from remote ranches and plantations in the Amazon. It turned out that most had been recruited in poor northeast Brazil with the promise of work. But they were held by armed guards and received no payment, because their employers said their debts had not been paid off. The workers had no way of knowing when, or if, the debts would be paid. Call it what you like, but that is slavery.

. . .

The world needs to fight a lot harder to prevent such trafficking. But one problem right now, say migration experts, is that stories of the traffickers color liberal perceptions of the rest of migration in ways that are bad news for those migrating because they want to. There is a pervasive liberal rhetoric about migrants as poor victims, forced to make journeys they do not want or would not make if they knew what lay at the other end. Behind this hand-wringing lies a mindset that is often as opposed to migrants as any xenophobe.

Hein de Haas of the International Migration Institute in Oxford condemns "apocalyptic views about a wave of desperate people fleeing poverty and warfare at home to enter the European El Dorado." These views led the UN's Office on Drugs and Crime to report that migrant smuggling "has become nothing more than a mechanism for robbing and murdering some of the poorest people of the world." In fact, the system is more deeply rooted, but also less criminal—and much more middle-class, says de Haas.

Few migrants are "fleeing." Nor are they the poor and destitute victims of evil snakeheads. Generally they are relatively well-off individuals making positive choices to improve their lives and those of their families. Their main problem is the extraordinary efforts made by receiving countries to prevent their arrival—often in the name of protecting the migrants.

Such mistaken ideas create mistaken policies. Most politicians in Europe believe that extreme poverty drives migrants, so if they promote economic development in poor countries, they will keep migrants at home. In his 2006 Bastille Day television address to the French people, President Jacques Chirac warned that "if we do not develop Africa, these people will flood the world." Two years later, following deaths on a boat full of Africans off the shores of Almería, Spanish Prime Minister Jose Luis Rodriguez Zapatero said, "As long as people are desperate and cannot feed their children, they will try to reach Europe."

The truth is the opposite. Most migrants today are not fleeing poverty so much as seeking wealth. They are aspirational. They come from middle-income countries like Mexico and the Philippines, Morocco and China. It is a critical difference, says de Haas. Economic development in poor countries usually encourages migration because it raises expectations, provides education, and gives people more get-up-and-go. And many do just that.

Another myth is that migrants are damaging their countries by leaving. In fact, says Stephen Castles, director of the International Migration Institute, migration is a normal and perhaps essential part of development for most countries. For one thing, migrants send home cash—about $280 billion of it a year, according to the World Bank. Indian families receive remittances worth $30 billion per year, Mexican families receive $26 billion, and Filipinos bank $19 billion.

This is just the cash logged through formal channels, like Western Union or travel agents. Other routes may contribute as much again. These include shared use of bank accounts, debit cards, and ATMs, and the informal courier systems known in Asia as *hundi*. Brenda Yeoh of the National University of Singapore says *hundiwalas*, often trusted locals from back home, "collect cash from migrants; use the cash to purchase goods like electronics; return to home country with goods; sell the goods in local markets; hand over cash to migrant's family and pocket the profit."

In recent years, migrants' remittances have been the largest source of foreign exchange for many countries. Tajik construction workers in Russia at one point contributed a staggering 45 percent of Tajikistan's gross domestic product. Moldova, Europe's poorest country, has 8 percent of its population working abroad, many as domestic servants in Italy and builders in Moscow. Their remittances reach half of Moldovan families and add 38 percent to GDP. Tonga, Lesotho, Lebanon, Honduras, Jordan, Haiti, and Jamaica all derive more than a fifth of their income this way. In some Bangladeshi villages, 80 percent of income comes from abroad.

Government development aid too often ends up in someone's Swiss bank account or dissipated in large bureaucracies. But this cash goes straight to families and local economies. It is spent on new houses, school textbooks, flush toilets, mosques, and dowries, or invested in small businesses and local trade.

Some argue that no matter how big these remittances from wealth-creating diasporas, losing the skills and energy of the best and brightest will doom poor countries. Yes, many expensively trained nurses, doctors, and other professionals are never seen again. But many migrants eventually return, bringing with them a wealth of knowledge and skills. India is currently gaining hugely from this return trade. At one point, twenty-five of the forty-five members of the Taiwanese cabinet had completed advanced degrees outside the country, according to Ronald Skeldon of the University of Sussex Centre for Migration Research.

But even migrants who never return provide a personal example of the value of getting educated and getting on, an example that will motivate many who do stay at home. The Philippines, for example, is the world's largest source of nurses. Hospitals and nursing homes in dozens of countries would fail without them. But for this reason, nursing is an extremely popular profession in that country. And in practice, most of the trained nurses never leave. As a result, the Philippines both exports lots of nurses and has more nurses at home per million inhabitants than, for instance, Austria.

Some people-exporting countries are so dysfunctional that they squander the potential benefits. But arguably, those countries are in such a bad way that we should think less about the impact of migration on their national statistics and more about the welfare of the families both at home and abroad. The answer, says Lant Pritchett, a free-market-loving economist at the John F. Kennedy School of Government at Harvard, is that generally the families do rather well—certainly compared with how they would have done otherwise. "Two of every five living Mexicans who have escaped poverty did so by leaving Mexico; for Haitians, it is four out of five."

Pritchett recalculated the wealth of nations to include the income of their diasporas. He found that Cubans, Lebanese, Somalis, and Fijians are 30 percent better off when that income is included; Haitians and Marshall Islanders are 40 percent wealthier; Liberians, 50 percent; Albanians, 60 percent; Jamaicans, 80 percent; Samoans, 90 percent; and Guyanese, 100 percent better off. The lesson Pritchett draws is that migration is good for families. And for many nations, mass migration is the best way to make their people better off.

So why doesn't it happen? In practice, most poor, developing countries are happy for their citizens to make a better life for themselves abroad. The restrictions are mostly from the richer, receiving countries. Pritchett calls these curbs a bigger discrimination against the poor than trade rules, the banks, or structural adjustment, or racial or sexual discrimination, or any of the other usual liberal targets. National border controls are the new apartheid of a globalized world economy. His bottom line, as a free-market economist, is that if capital is free to move round the world, then people should be too.

I'm not used to agreeing with free-market economists, even self-declared "compassionate libertarians." So I went to visit Pritchett in his tiny campus office at Harvard. Pritchett described his years at the World Bank, witnessing the results of this global apartheid—how a Vietnamese laborer could have nine times the purchasing power if she did an identical job in Japan; how a Kenyan could be seven times better off in Britain; how Tongan workers earn four times more in New Zealand and Mexicans seven times more in the United States.

"Construction jobs in the [Persian] Gulf have done more to bring wealth to ordinary Bangladeshi families than anything else in recent years." Sometimes, he agrees, the loss of skilled people will be damaging to the source countries. "But mostly that is from the movement of top people, the smartest and best educated. And they are the very people many Western countries are working hardest to attract. So the argument is disingenuous."

It is a damning analysis. The new global apartheid should be an affront to liberal consciences. Yet somehow we fail to be affronted. We come up with lots of reasons why, regrettably, the barriers have to remain. We pretend that even the keenest migrants are victims who would be better off back home. So we ban the movement of people to rich countries where they could work for decent wages, and instead "outsource" the jobs to countries where pay is rock bottom—like Vietnam and Kenya.

Except that some jobs can't be moved—jobs like picking grapes in California, cleaning toilets in Britain and hotel rooms in Italy, looking after babies in Washington, and gratifying the libidos of men from Tel Aviv to Toronto and Stockholm to Sydney. So we allow migrants to do these jobs, but either on short-term contracts or, most duplicitously, as illegal migrants—criminalized workers with no rights, conveniently ejectable at a moment's notice.

Such policies are inhumane and counterproductive. And here is a final irony. Erecting barriers to migrants frequently ensures that once people make it across the border (which they usually do), they never return. Take that most notorious border, the one between Mexico and the United States. In the past, half of the Mexicans who went to the United States to work returned home within a year. Some would go back and forth again and again. They were commuters. Now, of the estimated thirteen million Mexicans illegally in the United States, fewer than a quarter go home within a year. They can afford neither the smugglers' fees nor the risk of not getting back.

And, unable to return for conjugal visits, more and more migrants are paying smugglers to bring their wives and children across too. Some 35,000 Mexican children were deported in 2007 after trying to cross the border, half of them unaccompanied. Many more will have got through.

Tough U.S. border controls have also resulted in more people trying to cross at remote desert locations. The Buenos Aires national wildlife refuge in Arizona is being "slowly obliterated by Mexicans trudging north [along] 1300 miles of trails and 200 miles of outlaw roads," heading for the distant lights of Tucson, says Arizona writer Charles Bowden. But since the introduction of new, tough border controls, the death rate on the border has tripled, to more than five hundred deaths a year. Meanwhile, the proportion of would-be migrants who get caught is down from 30 to 5 percent. What sense does that make? The razor-wire fences, heavily armed border guards, and helicopters with searchlights may serve political purposes, but not much else. The dead are the victims of hypocrisy in the north.

CHAPTER 20

Footloose in Asia

I met Bridget in the Lucky Plaza shopping mall on Orchard Road in Singapore. Orchard Road is a haunt of tourists. But take the escalator past the glitzy electronics shops and logo-strewn clothes emporia to the upper floors of the mall, and you find a different world. First I passed the Ok Ka Lang pub. A bunch of girls in miniskirts rushed out with the kind of inviting "Hi" that you might hear in a million girlie bars across Southeast Asia. It was midafternoon, for heaven's sake. But beyond the bars, and a world away from their sleaze, the mall is where the maids from the tourist hotels hang out. There are Filipino and Indonesian cafés, parcel shippers, remittance arrangers, hairdressers, and cheap jewelry stores.

I found the office of the Humanitarian Organization for Migration Economics (HOME), which acts both as a "specialist Filipino and Indonesian domestic maid agency" and as a help point for foreign workers. Outside, Bridget was reading the messages in the window. She was from Luzon, an island in the Philippines, and was one of perhaps a quarter million Filipino maids and nurses in Singapore. She had been in the city for two years, doing housework and looking after a child for a professional couple. Now she was scouring the notice boards, looking for a job in Europe or North America.

In Singapore, one in seven households has a maid or other live-in help, and one in four of the workforce is a migrant. Most are from the Philippines. Why the Philippines? In Singapore, the law bans Chinese maids. They are "too like us, too much part of us. They might steal our husbands," one female employer told me. "Thai women are too sexy. Indonesian and Burmese women are popular, but not too many can speak English. There are Sri Lankans and Bangladeshis, too. But Filipinos usually have English as their first language. They are perfect. Many even have degrees, so they can make good tutors for our children."

With all those talents, Filipino maids often set their sights higher. There seemed to be plenty of demand: "200 maids needed in Abu Dhabi immediately; US$1000 a month," said one advertisement in HOME's office window. "US, UK, Australia: nurses required now," said another. Bridget wanted a place in England if she could get it. On the British minimum wage, she could earn in seventeen hours in London what she earns in a month in Singapore. The Lucky Plaza recruitment agencies will fix air tickets, visas, work permits, passport renewals, and all the other paperwork she would need before hopping on a plane to Heathrow.

The Filipino maid is much mythologized, says Brenda Yeoh of the National University of Singapore. She is variously seen "as a victim of forced labour and abuse, as a modern-day heroine whose toil in foreign lands sustains the Filipino economy, as a docile household servant, as a potentially sexually promiscuous 'other' woman in the family, as disposable labour or dutiful daughter or sacrificial mother."

Maids get mistreated, says John Gee, president of a Singapore-based NGO called Transient Workers Count Too. "Around half of domestic workers here do not get time off work throughout their two-year contracts. There is also a very patrician attitude here. Employers believe that if maids are given time off, they will get into bad company or get pregnant. The people who say these things often give more freedom to their children than they do to their maids." HOME runs a refuge for maids fleeing rape or physical abuse. It often has more than a hundred residents. While I was in Singapore, the refuge was facing closure after neighbors complained that the shelter affected the "ambience" of their neighborhood. But most maids settle for their lot, make their money, and—one day—go home.

Filipino maids have become famous worldwide. But they are just one part of a maze of cross-border migration paths that is remaking the human geography of East Asia, the most footloose region on earth. People here cross borders with even less fuss that the average citizen of the

European Union—for work and play, sex and shopping, sightseeing and golf, or just for the hell of it. Women are at least as much a part of the movement as men. Almost two-thirds of Filipino foreign workers are women. Across the region, increasing numbers of women no longer feel part of the traditional, enveloping family networks. They are giving up on marriage, and the repercussions extend far beyond the marriage bed and maternity ward.

For one thing, more and more Asian men are unhappy. They are failing to find the kind of economically, socially, and sexually subservient female that they often still crave. In South Korea, the effects of this on men are made worse by the fallout from the plague of sex-selective abortions during the 1980s that deprives them of adult female company today. Less marriageable men are in deep trouble. In Japan, more than a fifth of men with only basic secondary education fail to find a spouse, compared with only 6 percent of women with the same level of education.

Some men give up on the idea of marriage. Why have a wife when there are fast food outlets, labor-saving devices for the bachelor pad, and commercial sex services of all sorts? But others, particularly in richer countries like Japan, South Korea, and Singapore, are joining their womenfolk on the 747s to foreign lands. They go in search of the wives unavailable at home.

This is not a fringe activity for a few "loser men"; it is mainstream. A quarter of South Korean men now marry foreign women. Ethnic Koreans from China are popular, but so too are Vietnamese girls. Many South Korean men go bride hunting in Vietnam. But often they treat their foreign brides as little more than slaves—an issue that has become a scandal in both Korea and Vietnam recently. Similarly, one in five Taiwanese men now find a wife from the Chinese mainland, while others marry Cambodian, Indonesian, or Vietnamese women.

Hong Kong men are also seeking succor outside the territory. City blogger Kent Ewing summed it up: "The city's men prefer women from the mainland because they are more likely to follow traditional subservient norms of marriage. Hong Kong women are far better educated and more independent than their mainland counterparts and do not want to be bossed around by patriarchal husbands." Some Hong Kong demographers see mainland brides as a way out of the city's worsening birth dearth. The fertility rate among native women in Hong Kong is below one child per woman. Of 65,000 babies born in Hong Kong in 2006, almost half had mothers born on the mainland. Many Singapore men, too, go abroad in search of spouses. It would be incomplete not to mention here that Singapore's Gavin Jones, one of the region's most respected analysts of this phenomenon, has himself married a young Indonesian bride.

It is not just the young men who go abroad seeking wives. When unmarried old men from Asia's richer countries retire, they look for wives someplace where they can afford to live well, like Vietnam or Indonesia. Pity, then, the men of poorer Asian countries, squeezed between women departing for jobs abroad and rich foreigners snapping up what brides are available. In mainland China, the "little emperors" may have been loved by their mothers, but in adulthood they face what Nicholas Eberstadt of the American Enterprise Institute calls "an increasingly intense, and perhaps desperate, competition for the nation's limited supply of brides."

This transnational Asian male flight is not all about sex, or even a good woman's culinary and homemaking skills. For years, the better-heeled Japanese have flown south to soak up the sun, just as their northern European counterparts take off for Spain and Italy. Kuala Lumpur for a while ran an advertising campaign with the slogan "Malaysia My Second Home," designed to entice visitors. Penang Island is especially popular. In parts of Thailand there are whole rural communities full of Japanese retirees. The country also promotes itself as a center for international health care, with elderly Japanese as favored clients. The Japanese government once proposed building a whole town in Australia to accommodate its birds of flight.

One of the main things Japanese men want is golf. After Americans, they are the most ardent players. But Japan is simply too crowded to meet the demand for eighteen-hole courses. Clubs have long waiting lists and high fees. So instead, salarymen fly off at weekends to get in a few rounds at one of the hundreds of new courses being built in Thailand, Malaysia, Vietnam, Cambodia, and Indonesia. Some eventually figure they might as well go for good.

East Asia is becoming a global magnet for sun, sand, golf, and women. As the delights of the Mediterranean pall, Germans, Britons, and Scandinavians are joining the party in ever-increasing numbers. Sex tourism has been a mainstay of the Thai economy for a while now, with some districts of Bangkok, like Patpong, devoted to little else. Some of this sex trade is merely seedy, while some is brutal and obscene, with children sold for their sexual services. But increasingly, men are coming not just for sex but for partnership and marriage.

The British Embassy in Bangkok told the *Observer* newspaper in 2008 that of 800,000 British tourists who visit Thailand each year, 4,000 come to marry. Not everything goes well. The diplomats were giving background on how a retired engineer from the Ford motor plant at Dagenham in England came to be beaten to death by his wife, a forty-two-year-old former bar girl, and her lover in their shared home next to a rice paddy in a remote rainforest province of northeast Thailand.

Whatever the pitfalls, East and Southeast Asia are the new hot spots of international migration in the twenty-first century. And the region has an illustrious predecessor: the United States.

CHAPTER 21

God's Crucible

The United States of America is a nation of migrants. Americans sometimes forget this. But for a visiting European, it stands out. People in the Old World are used to migrants as a minority, but there is always a dominant "indigenous" nation that has spilt blood and buried its dead in the same soil for many centuries. Hence the fascination in coming to a country where, apart from the gracious presence of a scattering of Native Americans, the entire population goes back only a handful of generations. I love the New World. There is still something pioneering and "Wild West" about America, still something Southern plantation and Northern hunter. That can be bad—guns in the inner city, for instance. But it is invigorating too. Nothing is impossible in America, because so far everything has been possible. Cynicism has not yet taken root. America is still the biggest and greatest developing nation. It is a marvelous advertisement for the virtues of migration and a living condemnation of the world's xenophobes, racial purists, cultural imperialists, and ethnic cleansers.

In the late nineteenth century, the golden age of global migration, Portuguese whalers went to Rhode Island, Chinese "coolies" to San Francisco, and Japanese coal miners to Seattle. But for most, becoming an American meant first going to New York's Ellis Island, at the mouth of the Hudson River. After previous lives as a Dutch picnic spot and as Gibbet Island, where the English hung pirates, it became the main entry point for migrants to the United States from 1892 to 1954. More than a hundred million Americans are descended from the twelve million who passed through this factory for processing migrants. Today it is a museum, the most potent museum of America in a city fast becoming a theme park for the twentieth century, the American century.

America, as the museum declares, is "a multiethnic nation unparalleled in history." Ask Americans, as they do in the census, what their ancestry is, and you find almost a million who call themselves Africans, 180,000 Brazilians, 1.2 million Cubans, 3 million Dutch, 19 million English, 6 million French, 33 million Germans, 600,000 Haitians, 20 million Irish, 800,000 Japanese, 1 million Koreans, 300,000 Lebanese, 20 million Mexicans, 3 million Norwegians, 6 million Poles, 2 million Russians, 600,000 Salvadorians, 120,000 Turks, 600,000 Ukrainians, 1.1 million Vietnamese, 1.7 million West Indians, 300,000 Yugoslavians, and the odd Zairian. That's aside from the 35 million who describe their ancestry as African American, the 2.4 million who say they are American Indians, the 10 million who call themselves Asians, and so on.

In the seventeenth century, the majority of those who came were indentured white servants—the "white slave trade." Only after 1700 did black slaves make up the majority of arrivals. Some twelve million of them arrived in chains, brought by mostly English traders to serve the minority of white landed settlers. Many more died en route. (Turns out the Irish famine landlords at Skibbereen, the Freke family, who later took on the Carbery title, were running slave ships out of Bristol in England.)

In the nineteenth century, America began to fill with millions of Anglo-Saxons and Germans, Irish on "famine ships" after the potato blight, and Scandinavians fleeing their own poverty and famine. But the biggest influx followed the introduction of steamships running regular services. Northern Europeans came in vast numbers in the 1880s and 1890s, and after 1900, southern and eastern Europeans joined them. "America is God's crucible," wrote the Russian Jewish playwright Israel Zangwill in his play *The Melting Pot*, which opened in 1909. Arrivals peaked at 1,285,349 in 1907. All told, 26 million arrived, mostly through Ellis Island, before the First World War called an effective halt. This vast surge of humanity from Europe to North America a century ago remains the largest and quickest international migration of human beings the world has known.

The journeys were tough. It took up to fourteen days to cross the Atlantic aboard the overcrowded steamships from ports like Liverpool and Naples, Hamburg and Goteborg, Cherbourg and Riga, Antwerp and Palermo. They were usually once-and-for-all journeys. At Liverpool docks in England, migrants took a ball of string aboard, with the end held by a loved one on the dockside. As the boat set sail, hundreds of migrants paid out their string till the ends were left floating in the air.

On arrival, migrants who weren't staying in New York bought tickets for the new railroads. The train companies were as much in the business of selling land along their tracks as tickets to ride. Competition for settlers was intense. Nebraska offered "the best land in the West [with] the most liberal credit terms." California boasted of being "the cornucopia of the world . . . room for millions of immigrants . . . without cyclones or hurricanes." And when the best land started to run out, people came for jobs. Men worked in heavy industry and construction; women in sweatshops and as domestic servants. Scandinavians went to Minnesota, and Italians to New York City. Slavs sought out the steelworks, Finns the nation's timber yards; many Britons went down mines.

Americans were open and welcoming. They were all migrants themselves, after all. Well, they were welcoming up to a point. Hope was never far from fear, and the "keep out" sign never far from the welcome mat. African Americans had long known that the white northern Europeans' sense of openness and freedom had distinct bounds. The Chinese were the first newcomers to prove unpopular. Some 25,000 had taken the slow boat from the famine fields of Guangdong to join the California gold rush of the 1840s. They called the state Gold Mountain. They ran laundries, set up the first Chinese restaurants outside Asia, and later provided the labor that built the railways along which they ultimately spread throughout the United States.

But almost as soon as they arrived, the Chinese were attacked, and sometimes rounded up and thrown out. In 1862 the new California governor, Leland Stanford, took time out during his inaugural address to call for the expulsion of "the dregs" of Asia. (Though when his wife later contracted a dangerous lung disease, he still tracked down a Chinese doctor to cure her.) By 1882 there were some 300,000 Chinese in the United States. But the railroads were largely completed, and the country was entering recession. So Congress passed the Chinese Exclusion Act, and there were pogroms against the "yellow peril." Some Chinese headed for home, and the rest hunkered down in Chinatowns.

The backlash against newcomers hit southern and eastern Europeans after the 1917 Russian Revolution raised fears that Bolshevism might infiltrate America's nascent labor movement. Employers exploited a fear of foreigners. This was the time when the "science" of eugenics took hold. The Immigration Restriction League, founded in 1894 by students at Harvard, claimed science "proved" the inferiority of southern and eastern Europeans. Eugenicist Madison Grant was its vice president. Arrivals at Ellis Island were soon tested on their literacy and IQ. But when those of "inferior" races had the temerity to pass these tests, the Quota Act of 1921 simply limited arrivals by nationality.

Anglo-Saxons, Scandinavians, and Germans were welcome; other Europeans were severely restricted, and Africans and Asians all but banned. Hispanics were at first let in to work in the fields of the south and west, then thrown out during the Depression, recruited again during the Second World War, and thrown out again when the Korean War ended. Only in 1965 were quotas ended, and since then, Asians, Latin Americans, and Africans have come to dominate arrivals.

· · ·

But back to the golden age of migration and to New York City. I went to the Tenement Museum on the Lower East Side. The five-floor building, 97 Orchard Street, was built in 1863. It was the kind of place where the majority of migrants arriving in New York ended up. It had twenty family units, each of less than 325 square feet with a bedroom, kitchen, and parlor, plus two storefronts in the basement and toilets and taps out the back. Over the next seventy years, this one address housed seven thousand people from twenty nationalities. Early tenants included an actor, a cigar maker, a shooting gallery manager, a garment maker, a musician, and a peddler. To start with, this part of Lower East Side was known as Little Germany. By the turn of the century, Eastern European Jews had moved in. "Jewtown" was reputedly the most densely populated place on the planet, with up to a thousand people per acre (or roughly forty-three square feet each). Diseases like smallpox and tuberculosis (known locally as "the Jewish disease") killed half of all children.

But it was a hive of commercial activity. These overcrowded homes doubled as garment factories. New Yorkers made almost half the clothing for America then, mostly within a mile or so of Orchard Street. The term "sweatshop" was invented by social reformers here. I took with me to the Lower East Side *How The Other Half Lives*, published in 1890 by the campaigning photojournalist Jacob Riis. Riis described entering one tenement in Ludlow Street. "Up two flights of dark stairs, three, four with new smells of cabbage, of onions, of frying fish, on every landing, whirring sewing machines behind closed doors." The building housed thirty-six families, including fifty-eight infants and thirty-eight children over five—and a case of smallpox.

Workers for the sweatshops were always plentiful, said Riis. They would work for almost nothing. "Every ship-load from German ports brings them in droves, clamouring for work. Every fresh persecution of the Russian or Polish Jew on his native soil starts greater hordes hitherward. The curse of bigotry and ignorance reaches half-way across the world, to sow its bitter seed in fertile soil in the East Side tenements."

In 1890 one apartment on 97 Orchard Street was occupied by Harris and Jennie Levin. They were Jews from Plansk, near Warsaw, Poland, who crossed the ocean on their honeymoon. Harris was a "sweater," a subcontractor who had three employees working in the apartment for sixty hours or more a week. They made clothing from cut pieces of cloth, at seventy-five cents a dress. Sewing and finishing was done in the parlor and pressing in the kitchen using hot coals from the stove. Harris and Jennie's babies—successively Pauline, Herman, Max, Eva, and Fay—were often lying in a cot close by. The Levins moved out to Queens after a few years and continued their trade. Their descendants helped pay to re-create the apartment that immortalizes them and the world they inhabited. Israel may now have more Jews than the United States, but New York retains the largest Jewish population in any city in the world.

In the twentieth century, some 36 million people arrived and stayed legally in the United States. Add in the illegals, and you get to about 50 million. Only 12 million people left the country. Thanks to the recent surge, about 35 million of the 45 million Hispanics in the United States arrived, or were born, after 1990. In fact, two-thirds of the increase in the Hispanic population within the United States is down to births—making nonsense of the idea that closing the borders will stem the rise in their numbers anytime soon.

Hispanics make up a quarter of the workforce in Florida. But New York City is much more diverse. The city has constantly been reinvented by successive generations of immigrations. In the 1950s the government invited in Puerto Ricans. A fifth of the entire island took up the offer. The immigrant community and its battles with the locals formed the backdrop for the musical *West Side Story*. By 1964 Puerto Ricans made up a million of the city's ten-million population. Today there are more Puerto Ricans and their descendants in the United States than on the home island.

Then people from the Dominican Republic began to arrive. By 1969 there were more of them in New York City than anywhere except their own capital, Santo Domingo. By 2000 they were the city's biggest foreign-born population. More recently still, Ghanaians and Nigerians (forsaking their old association with Britain) and Senegalese (ditto with France) have arrived in large numbers. These groups revitalized the city after its 1970s crisis, when white residents left the bankrupt and crime-ridden city in huge numbers. Today New York's population is rising again. More than half the new births are to women not born in the United States—and crime is at a fraction of its former level. In suburbs of Queens like Elmhurst, Jackson Heights, Flushing, and Corona, most new occupants are migrants. Without them, New York would be a Rust Belt, has-been city with ghost-town neighborhoods.

The face of the city is constantly changing. In 1970 two-thirds of the people of New York were white non-Hispanics, with Irish, Italians, Poles, Jews, and Greeks dominating many neighborhoods and professions. Today, whites dominate only among those over sixty-five. The ultraorthodox Jews of Williamsburg, with their large families, prosper. Italians remain strong on Staten Island, and the Greeks in Astoria. But the guidebooks are mostly out of date in their ethnic geography. Most of them don't even notice the Ecuadorians, Bangladeshis, Jamaicans, Pakistanis, and many other groups, all recently established for the first time.

Chinatown is still here, of course, on the Lower East Side, just where it was a century ago, with its crowded tenements, fire escapes, and basements, its ginseng, noodles, and dumplings. There is a funeral home next to a karaoke lounge next to a basement laundry. In a basement in Bayard Street, a company called Noom Conruction [*sic*] sells air-conditioning systems. Close by is Foot Long Life Inc. (massage at a dollar a minute), the Restaurant Workers Association, the Nice One bakery, the Confucius Plaza Housing Project, and a table-tennis training area.

But even here, there are signs of change. The Cantonese are being crowded out by the Fujianese. Thai, Vietnamese, Korean, and Malaysian restaurants are muscling in. Tony Bennett booms out of Giovanni's outsize speakers in neighboring Little Italy. The old Museum of Chinese America is now the Chinatown Senior Citizens Center and doubles as a nuclear fallout shelter.

Take the subway and soon, three stops on the Q train from Coney Island, you are down on the boardwalk at Brighton Beach. There I heard nothing but Russian and Hebrew spoken. The neighborhood, named after the quintessential English seaside town, became filled with Jewish concentration camp survivors in the 1950s and more recently with émigrés from Russia and Ukraine. Today it is widely known as Little Odessa. The signs for dentists and doctors, lawyers and accountants, all bear Jewish names. Well, except the ones in Cyrillic script, which I couldn't decipher. I could have bought a "Moscow Breakfast" at "Russia on the Beach" (panini and hot dogs also available), challenged Russians playing chess on a bench, joined a pair of drunken Jews singing their prayers overlooking the ocean, listened to the Slavic violin player at the Tatiana Grill, eaten "Russian-style ravioli" at the *New York Times*–recommended Glechik café, or noshed on a potato knish made at Coney Island.

On the main street, beneath the overhead train line, there was a shop that stocked nothing but vodka, and another full of Moscow Dynamo soccer shirts and Russian knickknacks. The supermarket offered Slavutich beer from Kiev and Baltika from St. Petersburg, Europe's second largest brewery, and Urquil from the Czech Republic, along with Armenian brandy and shelves heavy with caviar. I settled down inside the Primorski restaurant, which specializes in Ukrainian borscht, Georgian dumplings (*khinkali*) and stew (*solyanka*), and my choice, a Georgian sausage called *kupaty*. The band was tuning up and the waiters were putting on their ties, ready for Saturday night showtime.

I read my notes. New York City today has three million first-generation immigrants, more than at any other time in its history. It remains the starting point for most people arriving in the United States, just as it was in the days of Ellis Island. Nearly half the population speaks a language other than English at home. New York, New York. It's a wonderful town still. It is a genuine city of the world. Only London is up there with it. This, I feel, is the future of our planet. Bring it on.

. . .

The end of the immigration quotas in 1965 revived migration into the United States. By the start of the twenty-first century, numbers arriving were approaching those at the start of the twentieth century. Many were welcomed, but not all. Sometimes the difference in acceptability was political. Uncle Sam welcomed escapees from Castro's Cuba but not those from Haiti, who were regarded as the victims of their own failures rather than of communism.

But where a century ago some newcomers were despised for a supposed eugenic inferiority, the parallel concern in recent times has been environmental. Many American environmentalists have actively campaigned to stop Hispanic migrants in particular from reaching the United States, on the grounds that their large numbers will damage the U.S. environment. The Carrying Capacity Network sounds like a benign environmental organization. But beneath its mission statement to "secure the sustainable future of the United States," its main campaign is to halt migration to the United States and oppose amnesty for illegal aliens. Typical headlines on its Web site are "New illegal alien amnesty threat" and "Only increased border control can prevent forest fires." I couldn't find anything actually discussing carrying capacity as a concept, nor anything addressing the fact that population density in the United States is much lower than in most of the countries the immigrants come from. The United States has 80 people per square mile, compared with 140 in Mexico, 475 in the Dominican Republic, and 760 in Haiti.

The Carrying Capacity Network has some influential directors and advisers. They include Tom Lovejoy, a biologist and former vice president of the World Wildlife Fund in the United States; green economist Herman Daly; entomologist and agriculturalist David Pimentel; the inventor of the concept of the ecological footprint, William E. Rees; and Virginia Abernethy, a prominent anthropologist and avowed "ethnic separatist." Abernethy gave a keynote address at the 2004 annual meeting of the Council for Conservative Citizens, a white supremacist organization that describes itself as "advocating against minorities and racial integration."

Many of these individuals, most prominently Pimentel, were part of an anti-immigration campaign organized within the Sierra Club, the United States' oldest, largest, and most influential environmental group. In 2004 the anti-immigration faction put up a slate for election to the club's board of directors, under the banner "Sierrans for U.S. Population Stabilization." Nasty stuff, in my view. I am glad they lost.

. . .

Most American migrants come to New York. But not all. Most Indians are in California, most Dutch and Finns and Arabs in Michigan, most Belgians in Wisconsin, most Bolivians in Virginia, most Bulgarians and Poles and Lithuanians in Illinois, most Croats and Slovaks and Syrians and Welsh in Pennsylvania, most Czechs in Texas, most French and Vietnamese in Louisiana, most Germans and Hungarians in Ohio, most Norwegians in Minnesota, most Portuguese in Massachusetts, most Cambodians in Kansas, and most Cape Verdeans and Liberians in Rhode Island. Recently, Sudanese migrants have headed for Nebraska ("the best land in the west"), Liberians for Minnesota, and Nigerians for North Carolina.

Oh, and some forty thousand Somalis have ended up in Columbus, Ohio. That intrigued me. The unassuming state capital of Ohio is, on the face of it, quintessential middle America. During my visit, the city was completely obsessed with a weekend college football game. "Why Columbus?" I asked Hasan Omar, the tall, slight, and gentlemanly head of the Somali Community Association on the north side of the city. It was hard to think of two places more different than Ohio and the deserts of Somalia.

The arrivals began in 1995, he said, when the United States began to admit refugees from the Somali civil war, who had been living in huge camps at Dadaab in the Kenyan desert, close to the border. The camp inmates were togged up in gray sweatshirts and yellow shawls by the U.S. refugee service and put on planes from Nairobi to reception centers in the United States.

On arrival, they had some false starts. Many were put in bad neighborhoods in Atlanta and left en masse for Lewiston, Maine, where the mayor called for them to be kicked out. Others clashed with Native Americans in North Carolina and with Hispanic meat packers in Grand Island, Nebraska. A couple of Somali families found themselves in Columbus. It seemed good. "Word spread that there were a lot of warehouses and factories here that needed labor, and you could get a two-bedroom apartment for a hundred dollars a month. People just came." Soon the new arrivals were phoning their families in the camps back in the desert: once you get a ticket on the plane to the United States, Columbus is the place to head for.

Back home, the Somalis are organized into clans. Clan squabbles lie behind a lot of the fighting that has wrecked the country in recent years. And there is a racial divide between the original Somalis and the minority Bantu groups—like the Mushunguli people—who were imported as slaves from Mozambique and Tanzania and later became farmers. Discrimination continues. The Bantu are still "less used to modern ways," said Hasan. "Some of them don't know how to turn a light bulb on." Non-Bantu Somalis spend an average of five years in the refugee camps before getting their exit papers; for Bantu Somalis, the average is twelve years.

Hasan has been in the United States for longer than most Somalis. He came first as a student. He founded the Columbus Somali Community Association in 1998. "We came from a war-torn country and didn't trust each other. But we don't have clans here. We try to live as one community, as Americans," he told me over tea and samosas in a Somali-owned café downstairs from his office. "We are political refugees rather than economic refugees. We are not destitute. We have skills: we are engineers, economists, and teachers. Our mentality is that this is an opportunity. We have two thousand of our youngsters at the state university and other colleges here in town. My son enrolled in college just last week."

They are in business too. They run shopping malls and car dealerships and a health center. There are four hundred Somali-owned businesses in town. In addition, about a thousand Somalis drive taxis. "In Europe, there is too much welfare for refugees," Hasan said. "Here in America there is less welfare, but there are more job opportunities." Some six hundred Somalis work at Limited Brands, packaging Victoria's Secret lingerie. Other big local employers that take on Somalis include Wendy's, Red Roof Inns, Gap stores, and JCPenney. Somalis are good fishermen. A bunch of them had recently left their families in Columbus and headed to Alaska to join local fleets. "They will do it for a year, save their money, and come here to start a business," Hasan said.

After 9/11, the people of Columbus got worried about a large crowd of African Muslims in their midst. There were fights between Somali and African American children in schools. "We decided to spread out a bit," Hasan said. "We realized that if you stay in one place, you get a ghetto mentality and opportunities are more limited." Nonetheless, he said, "for me, Columbus is the most successful Somali community in the world." They have their own three-hour radio program on the local FM station on weekends. The producer dropped by as we talked. About fifteen thousand Columbus Somalis have become American citizens so far. Hasan expected them all to vote in the 2008 elections. "We have invited local candidates to come and talk to us. I think we'll be running for office ourselves soon."

Later I asked a refugee volunteer at the university about Somali women. They aren't quite living the American dream. Most, she said, are still circumcised. And they have formed organizations to fight abuses like rape. "They still often won't talk without their husband being in the room. But they are working more and getting more power. In some ways they do well here, because they will do anything to work, while some of the men, who were big men in Somalia but have lost status here, are too proud."

Columbus could be Anytime, Anywhere, USA. The United States today, as throughout its history, is a made-up country of migrants. The old German village in Columbus's city center is now a tourist spot with no sign of any Germans. But look further, and you'll find Koreans running dry cleaners, Indians in charge of motels, Cambodians trading diamonds—and the Somalis running malls. Now, too, there are communities of Ghanaians and Ethiopians, of Hmong tribals from Southeast Asia and Rwandans from central Africa. There are even newer arrivals from Burma and Bhutan, and Iraqi refugees from Jordan. Suddenly the Somalis are no longer the new kids on the block. They are old hands.

My cab back from Hasan's office was driven by an Eritrean called Abraham, who worked for the Irish-owned Shamrock Taxi company. In town, the Columbus Italian Festival was starting. This is America, I thought. This is the new world. We should not fear it.

PART SIX

·

REACHING THE LIMITS

Is a large population good or bad for rural communities, for cities, for countries, or for the planet? Will sheer numbers overwhelm our ability to feed the world? Or do more people also create "more hands to work and brains to think"? It is a question that has been asked ever since Malthus. Some say we now face a new threat: a "youth bulge" in which unemployed and alienated baby boomers in developing countries create an anarchy in which terrorists and criminals rule. Is Malthus about to be proved right after all? Or will the peoplequake spur us to a greener way of living?

CHAPTER 22

The Tigers and the Bulge

First the good news. Sometimes having more people is good for business. Take Japan. Sixty years ago it was a basket case: the "most over-populated country there has ever been," according to UNESCO's Julian Huxley, and the place that "cannot possibly feed itself," according to William Vogt in *Road to Survival*. Yet even as Malthusian doomsters wrote Japan's epitaph, the country's economy was reviving in workshops across the country.

The Sony Corporation began as a radio repair shop in a bombed-out building in Tokyo in 1945. Sanyo set up shop the following year, making bicycle lamps, then radios and washing machines. Honda assembled its first motorcycles in 1948. Soon Japan was selling these goods to the world. The Land of the Rising Sun lived up to its nickname. With GDP rising by more than 10 percent a year, Japan could soon afford to import all the food it wanted. Huxley, Vogt, and the rest had been proved spectacularly wrong. Japan became the original "tiger economy" of Asia and was, for a time, the richest nation in the world.

Taiwan, Hong Kong, and Singapore followed. South Korea, too, after the end of the Korean War. Development economists in the 1950s had taken a dim view of the prospects of all these places. I remember being told by school geography teachers, even in the late 1960s, that all four were doomed because, like Japan, they lacked both land to grow their food and natural resources such as metals or oil to trade for food. Worse, their populations were growing fast.

But the prognosis was wrong. In Korea, the Hyundai car company was founded in 1946, and the company that became LG Corporation began making chemicals and electronics the following year. Singapore became Asia's number one port, Taiwan was an electronics manufacturing hub, and Hong Kong made pretty much everything. From 1965 to 1990, all four countries joined Japan to make up the five fastest-growing economies in the world. What made the tigers roar? They certainly invested well. But researchers in recent years have flagged up demography as another critical—perhaps the critical—factor. Japan, South Korea, Singapore, Taiwan, and Hong Kong created societies in which, during the critical years of hypergrowth, the overwhelming majority of their populations were young, educated adults, ready and willing to work.

The process of creating this demographic engine of growth had two phases. First, each country had a postwar baby boom. Nobody planned it; it just happened. As the boomers hit the workforce in the 1960s, they stimulated an already growing economy. Meanwhile, fertility crashed. The boomers were so busy working that they had many fewer children than their parents. So the number of children who had to be supported and educated by this large workforce fell to very low levels. The countries could invest more in factories and less in schools. Bingo.

Across East Asia, between 1965 and 1990, the working-age population grew four times faster than the "dependent" population of children and the old. From the north shore of Hokkaido to the tropical waters around Singapore, more than two-thirds of the population was of working age. And working. They were also getting wealthy. Incomes rose by an average of 6 percent a year. Demographers talk about this phenomenon as a "demographic window." As countries move from high to low fertility, they experience a period of a couple of decades when demographic conditions for rapid economic growth are nearly perfect.

Many countries have seen their economies burst into life during this phase of their demographic development—most recently, China. In the 1960s and early 1970s, Mao Zedong engineered a huge baby boom that saw the country's population reach a billion in 1980. His successors threw the gears into reverse, culminating in the one-child policy. Nobody quite realized it at the time, but the combination of the two policies created a demographic sweet spot. Like its neighbors, China had a large pool of working adults and a relatively small number of dependents. The young adults formed a vast army of 140 million migrants that moved from the countryside to the new industrial cities. And the Chinese economy reaped the benefits, with double-digit economic growth that has transformed it into the new workshop of the world.

Others Asian countries are now trying to repeat the trick. Thailand's fertility has crashed in forty years, from 6.6 children per woman to 1.6. Its economy has been growing by 5 percent a year. Vietnam has also had a population surge followed by a fertility decline, to 1.9 children per woman. As its schools empty, the majority of its 85 million people are now of working age, and its economy has been growing by more than 6 percent a year. Ho Chi Minh City is currently one of the fastest-growing cities in the world.

Who's next? A lot of money is going on India. It has doubled its population in forty years. But fertility is now falling fast. And the economy is stirring. Austrian demographer Wolfgang Lutz says that "the next twenty-five years offer India its chance. The proportion of children in the population will fall by a third, while the proportion of elderly will remain small."

Indian economist Sanjeev Sanyal says it's now or never for his country. Right now he works in Singapore, the quintessential tiger economy. We met in his office on the seventeenth floor of the Deutsche Bank's palatial offices on Raffles Quay, one of the most prestigious addresses in Asia. We looked out across the city-state's huge container port and watched builders erecting new gleaming towers right next to us. One day, Sanyal said, his country of a billion people could aspire to the wealth of Singapore.

"India has been going downhill for a thousand years, but my generation of Indians is experiencing the revival," he said. "We have the demographic window, and we have the capital savings. We are ready to go." He handed me a copy of his new book, *The Indian Renaissance*, in which he argues that if India opens its borders to market capitalism, it can deploy its demographic might to take the lead spot in the world's economy. "India can do to China what China did to Japan," Sanyal said. "We will take over the top spot between 2025 and 2045."

Will it happen? Lutz offers a caution. To reap the benefits, India will have to invest in its people. The original tiger economies were driven not just by a large workforce, but by a healthy, literate, numerate, and numerous workforce. "It's this human capital, not crude numbers, that correlates best with economic growth," he says. And that presents India with a problem. India may have a sizeable elite, university-educated middle class—people like Sanyal—but it also has 450 million illiterates. When China began its rapid growth, three-quarters of the working-age population had at least secondary education. Ran Bhagat, of the International Institute for Population Sciences in Mumbai, agrees: "Our problem remains illiteracy. Only if we can do something about that quickly will we reap the benefits of the demographic window."

Lutz says a single number can measure whether countries are ready to exploit the demographic window. He calls it "literate life expectancy" —the number of years during which a country's average citizen is literate. In most European countries and the earlier Asian tiger economies, where literacy is nearly universal from childhood and lives are long, the literate life expectancy is sixty-five years or more. At the bottom is Afghanistan, with fourteen years. China's literate life expectancy is healthy at around fifty. But India's is still not much above thirty. It is far too little, Lutz says. It could undermine India's economic ambitions.

Behind India sit the numerous countries of sub-Saharan Africa— real basket cases, according to today's economists. Is there any chance of them benefiting from a demographic window of opportunity? Could be. Africa has done the first part of the job: creating a huge youthful population. It has the youngest population of any continent in the world, with around 44 percent of inhabitants under age fifteen. Those kids are an economic drag now. But soon they will be of working age. So it is time for the next phase, cutting numbers in the next generation.

The good news is that a surprising number of African countries now have falling fertility. As recently as 2002, twenty African countries had an average of six children per woman. But by 2008, only nine did. Several—including Ghana, South Africa, Zimbabwe, Botswana, and Kenya—have seen their fertility fall by more than a third. So the region can expect a fast-rising ratio of young adults in its population in the next two decades. Lori Ashford at the Population Reference Bureau predicts that Ghana's working-age population will peak at around 65 percent during the 2030s, Namibia's will peak in the 2040s, and those of Ethiopia, Uganda, Kenya, and others soon after. That will be their opportunity, if they can grasp it.

But that remains an unlikely prospect while Africans are so poorly educated. The region's literate life expectancy remains lower even than India's, averaging twenty-four years. But if they could get all children through primary education soon, then the figure would be above thirty-five years by 2030. That would still be the lowest in the world, but it might be enough. Don't write off Africa.

· · ·

Not all countries succeed in using the demographic window to send their economies into industrialized overdrive. Conditions have been right for economic takeoff in Sri Lanka and Pakistan, for instance. But both have instead descended into political chaos. Sri Lanka has been at war with its Tamil minority, and Pakistan has been wrecked by corruption and the Taliban tide. Some say the political chaos simply prevented them taking advantage of the window. Others argue that the surge of young adults in the populations of these countries became the cause of the chaos. Those who would have fueled economic transformation instead triggered social disintegration.

Some demographers call this phenomenon the "youth bulge." The risk is of an alienated, unemployed generation of youths bringing terrorism and political breakdown in their wake. There is growing concern about this youth bulge, especially in an era of reduced economic growth. It is the ugly flip side of the same demographic phenomenon that created the tiger economies.

My introduction to the youth bulge came in a report published by Population Action International in 2005. PAI is the successor body to General William H. Draper's Population Crisis Committee. The report, *The Security Demographic*, was dedicated to the memory of the man who brought the population explosion to the heart of cold-war policy making in the United States in the 1960s. It says that our post-9/11 world is awash with urbanized, radicalized youths with no jobs—and such people often start wars.

This was not just rhetoric. The report had some research backing. Its authors found that countries in which those under thirty make up more than 40 percent of the adult population are more than twice as likely "to experience an outbreak of civil conflict." They called this relationship "striking and consistent." They said the link shows up especially in the fast-growing cities of modern Africa, the Middle East, and South Asia, where slums and shanties breed violence. "The risks of deadly violence between governments and non-state insurgents, or between state factions, that are generated by demographic factors may be much more significant than generally recognised." The report predicted future conflicts driven by youth bulges in Bangladesh, Nepal, Laos, Bhutan, and the whole of eastern Africa.

One of the authors, Richard Cincotta, says the world faces a new "demographics of insurgency, ethnic conflict, terrorism and state-sponsored violence. The vast majority of [terrorist] recruits are young men, most of them out of school and out of work. It is a formula that hardly varies, whether in the scattered hideouts of Al Qaeda, on the back-streets of Baghdad or Port-au-Prince [in Haiti] or in the rugged mountains of Macedonia, Chechnya, Afghanistan or eastern Colombia."

There is today a pronounced youth bulge across the Muslim world. Of the twenty-seven countries with the largest proportions of young adults in their populations, half are Muslim. Between 1995 and 2005, the percentage of young adult males in largely Muslim countries like Pakistan, Iran, and Saudi Arabia rose a phenomenal 26 percent. Few of these countries are using the high numbers of young adults as an economic driver. In most of them, youth unemployment is very high. And so is disaffection. In India, as we have seen in chapter 16, the abortion of female fetuses has created a male youth bulge that former UN demographer Joseph Chamie warns may lead to "bachelor gangs creating crime and disorder."

This is scary stuff. Jeffrey Sachs, Columbia University's development guru who counsels many world leaders, says that "extremely young populations, with a high proportion of young men" are "probably an important factor in the crises of governance throughout sub-Saharan Africa and the Middle East, including violence and terrorism." He notes that "in Saudi Arabia, home of most of the 9/11 hijackers," there are more individuals under twenty-one than over, "representing a worrying lack of adult supervision."

Gunnar Heinsohn, a genocide researcher at the University of Bremen, says youth bulges can help explain everything from twentieth-century European fascism to slaughter in Darfur, from civil wars in Lebanon and Algeria to Palestinian uprisings against Israel. "Young men tend to eliminate each other or get killed in aggressive wars until a balance is

reached between their ambitions and the number of acceptable posi-
tions available in their society." As we saw in chapter 17, some Bedouin
academics in the Negev Desert in Israel predict insurrection among
their own disaffected youth bulge.

The political repercussions of these youth bulges are not necessarily fundamentalist or violent. The height of the European and North
American youth bulge in the 1960s saw left-leaning radicalism and the
rise of pop culture—the hippie generation. And in Iran today, the same
demography is generating a reformist reaction against the clerics.

Nonetheless, some environmentalists see a new nexus of disorder where youth bulges, worsening environmental problems, and conflict feed off each other. The leading exponent of this idea, Thomas
Homer-Dixon of the University of Waterloo in Canada, says, "As environmental degradation proceeds, the size of potential social disruption
increases. . . . It will have incredible security implications." This analysis
has been influential. The U.S. National Security Strategy warned as long
ago as 1996 that "large-scale environmental degradation, exacerbated by
rapid population growth, threatens to undermine political stability in
many countries and regions."

Robert Kaplan, in his influential 1994 essay "The Coming Anarchy," painted an alarming picture of dark, lawless cities, bush shanty
towns, and refugee camps occupied by stateless drug cartels and growing bands of refugees and would-be terrorists, ruled by "the environment as a hostile power." He predicted that "the violent youth culture of
the Gaza shantytowns may be indicative of the coming era." He wasn't
wrong about Gaza. The small Palestinian enclave, trapped on a strip of
desert between Israel and the ocean, is today one of the most densely
populated, environmentally damaged, and violent places on the planet.
It also has one of the youngest populations. After the Israeli invasion of
Gaza in early 2009, Heinsohn blamed a youth bulge for the insurrection
that prompted Israeli action. "Nearly half the people are under fifteen
and the terrorists hide behind them."

Heinsohn's most original twist, however, was to blame Western
and UN "largesse" to the imprisoned people of Gaza for perpetuating
the bulge. In an analysis right out of Malthus's assaults on English Poor
Laws, he argued that "a large majority of Gaza's population does not
have to provide for its offspring. Most babies are fed, clothed, vacci-

nated and educated by the UN.... One result of this unlimited welfare is an endless population boom. Unrestrained by such necessities as having to earn a living, the young have plenty of time on their hands for digging tunnels, smuggling, assembling missiles and firing them at Israel." He predicted Gaza's population would double to three million by 2040, a population explosion he called "Gaza's extreme demographic armament."

Such analysis has a basis in demographic reality. But it is crude demographic determinism that ignores the politics. Yes, Gaza has a large supply of unemployed, alienated, and angry young men and women ready to take up arms. Yes, its high fertility is adding to the number. But it is Israel's economic blockade of Gaza—rather than Western aid—that has shut down economic activity and left the youth bulge with little possibility of earning a living and every incentive for assembling missiles. And Gaza's continuing high fertility (in contrast to falling fertility on the West Bank) is rooted in its siege situation. Heinsohn's analysis serves, like Malthusianism in Ireland in the 1840s, as a self-fulfilling prophesy. Demographics can too often be a handmaiden for the politics of fear.

Japan in the 1960s and Gaza today represent the two extremes of what happens when countries experience a youth bulge. Either a country's youths are harnessed and its economy can prosper, or they are not, the national economy stagnates, and it risks exploding in violence. Demography drives both, but what ultimately happens has as much to do with politics as demographics.

CHAPTER 23

Footprints on a Finite Planet

Humans are the most successful species in the history of our planet. But our very success has encouraged us to increase numbers so much that our own survival is now at risk. The danger is of no more tiger economies but a lot more Gazas.

Scientists have long pondered how many people the planet can sustain. In 1679 a Dutchman called Antoni van Leeuwenhoek, an early microbiologist and friend of the painter Johannes Vermeer, calculated that the inhabited area of the planet was some thirteen thousand times larger than Holland, which then had a population of about a million people. So he reckoned the world could handle at least thirteen billion people. Current environmentalists have somewhat lower figures.

Paul Ehrlich figured that what he called the planet's "carrying capacity" might be about five billion. More recently, scientist have calculated that with 6.8 billion people on the planet, we were by 2008 consuming 30 percent more resources each year than the planet produces, so we are trashing the rainforests, emptying the oceans, eroding the soils, and filling the air with greenhouse gases. In the jargon, we are "drawing down" our natural capital. They suggest that the way we live now, the true long-term carrying capacity of the planet is about 5.2 billion people.

There are other, lower figures. Britain's leading organization campaigning for smaller populations, the Optimum Population Trust, which counts Ehrlich among its patrons, says we have to bring our numbers down to three billion or face "nature's brutal population policies . . . increasing the death rate by famines and disease." The scientist James Lovelock, inventor of the Gaia hypothesis, says we are treating the planet so badly that we are likely to require a population crash to about one billion people before the world can again live within its ecological means.

It is easy to see why there is so much pessimism. We are wrecking many of the earth's life support systems. An inventory of the state of nature's bounty makes a sobering read. We have removed half the planet's forests. They once covered two-thirds of the planet's land surface but are now down to one-third. We have destroyed about a quarter of its topsoil through plowing and erosion. We have killed off most of its large animals and probably nine-tenths of its fish stocks. We are consuming about 40 percent of the plant matter grown on the planet and diverting about 60 percent of its river flows for irrigation, for cities and industry, or to hydroelectric reservoirs.

Every year we extract from underground, and then burn, the fossilized remains of plants that it took nature about a million years to produce. The resulting carbon dioxide emissions have so far warmed the atmosphere by one degree. As a result, we are melting the Arctic, raising sea levels, and intensifying droughts, floods, and storms. The warming will last so long that we are probably close to canceling the next ice age.

In addition, we have altered the chemistry of the atmosphere sufficiently to rip a hole in the protective ozone layer and to acidify the rain and even the oceans. We are now the dominant force in the nitrogen cycle. Fertilizers are flushing so much of the element into soils and water sources that forests die and giant dead zones form in rivers, lakes, and oceans. We have created (but thankfully have so far refrained from using) hydrogen bombs with the ability to wipe out most life on earth. This is a planetary crisis. And many people blame it on human numbers.

Our sheer numbers have clearly been crucial to what has happened, but they are only part of the story. In *The Population Bomb*, Paul Ehrlich pointed out that our environmental impact on the planet is a combination of three things: the number of individuals, the consumption of each individual, and the resources needed, or pollution created, in satisfying that consumption. He argued that population growth was the dominant factor in increasing our environmental impact in the 1960s. Perhaps for a while it was. But since then, the pace of population growth has slowed. And what growth continues is increasingly confined to the planet's poorest people, those who consume the least. So the impact of those extra people is surprisingly small. By late in the century, when some threats like climate change are expected to become most serious, our population may be falling.

If population was the only thing we had to worry about, we might be okay. The trouble is that, as population growth has slackened, Ehrlich's second factor in humanity's impact on the planet has come to the fore. Rising consumption is now a much bigger threat to the planet. It is responsible for almost all our increased ecological footprint in the past thirty years, the period when analysts say we have overshot the planet's carrying capacity. Despite the rise of economies like China's, that increase in consumption is mainly in the rich world, among those already consuming the most.

The average citizen of the United States has an ecological footprint of 23.5 acres—that is, the amount of the planet's surface needed to provide him or her with food, clothing, and other consumables, and to soak up his or her pollution. Meanwhile, Australians and Canadians require 17 acres, Europeans and Japanese 10 to 12 acres, the Chinese 5 acres, and Indians and most Africans less than 2.5 acres. Of course, there are rich people in the poor world and vice versa. But if we look just at the richest billion people on the planet, their average consumption of resources and production of waste today is thirty-two times that of the average for the nearly six billion remaining people.

A separate calculation has been done on who is responsible for the greenhouse gas emissions behind climate change. It turns out that the poorest three billion or so people on the planet (roughly 45 percent of the total) are currently responsible for only 7 percent of emissions, while the richest 7 percent (about half a billion people) are responsible for 50 percent of emissions. A woman in rural Ethiopia can have ten children and her family will still do far less damage, and consume fewer resources, than the family of the average soccer mom in Minnesota or Manchester or Munich. In fact, in the unlikely event that her ten children live to adulthood and all have ten children of their own, the entire clan of more than a hundred will still be emitting only about as much carbon dioxide each year as you or I will.

So to suggest, as some do, that the real threat to the planet arises from too many children in Ethiopia, or rice-growing Bangladeshis on the Ganges delta, or Quechuan alpaca herders in the Andes, or cow-pea farmers on the edge of the Sahara, or chai wallahs in Mumbai, is both preposterous and dangerous. This is not to say that population is irrelevant. The quadrupling of global population during the twentieth century helped bring us to the edge of the abyss. But any analysis of the damage being caused today by rising population and rising consumption must conclude that consumption is the greater peril.

It is of course true that poor people with small ecological footprints may grow rich, or produce children who grow rich, eventually assuming ecological footprints as great as ours. If they do that, it is hard to see anything other than disaster ahead. If nothing else, climate change will create such chaos that the prospects of feeding five billion, let alone more, will be all but impossible. But there is good news. We can reduce our ecological footprints while keeping, if not every aspect of our lifestyles, then at any rate those parts of our lifestyles that make our lives truly worth living.

. . .

The wonder is that we rich-world consumers have come so far without already precipitating a major crisis of the kind envisaged by Malthus, Vogt, Ehrlich, and others. Our bacon has been saved by the third, and least discussed, element in Ehrlich's equation. As our technologies improve and become more efficient, so we are getting smarter at generating wealth. We are using fewer resources and creating less waste in making each dollar. Our power stations produce more power from the same amount of fuel, our industries take less power and make better use of metals, we replace scarce materials with more abundant ones, we recycle more things we used to landfill, and so on.

The gains are remarkable. The trouble is that they are masked by our rising consumption. The most obvious example is our cars. They are much more fuel efficient and much less polluting than vehicles of the same weight and performance even a few years ago. Unfortunately, we insist on taking advantage of that fact by driving bigger vehicles, like SUVs, and driving them longer distances. So the gains are lost. We are still using more resources.

The hope now must be that our growing concern for the planet's limits will drive us to cut the environmental impact of generating the good things in life and to be more green-minded in our sense of what the good things in life are. Chris Goodall, in his book *How to Live a Low-carbon Life*, concluded that most of us could reduce our carbon footprints by 75 percent at little inconvenience. So there is a way forward that requires neither culling large numbers of people nor sacrificing our quality of life. We just have to take it.

We have to trust that Ester Boserup, the Danish economist, was right when she said that crises, and our awareness of them, are what stimulate major innovation, both in technologies and in the way we organize things. I believe we can. We humans are good problem solvers, once we have figured out what the problem is.

My optimism won't necessarily be borne out by events. Humans don't always get things right. The planet is littered with the remains of past civilizations that crashed painfully, often as their natural environments gave out. And this time we have a global civilization that is impacting on the planet as a whole, especially through climate change. We have to get this one right. I can see why many reckon there is little chance of success and why, metaphorically at least, they are ready to head for the hills. But I am not.

There are plenty of practical examples to justify my optimism that Ehrlich's third factor can come to our rescue. The green revolution happened because we recognized a crisis over the planet's ability to feed a population expected to double in a generation. Faced with that crisis, we acted. Both Europe and North America have dramatically improved their domestic environments in the past half century—reducing smog, cleaning rivers, and reforesting. A handful of poor countries are now taking the same road, reversing the loss of tropical rainforests, for instance.

Take Costa Rica. As ranchers and loggers ransacked the country's forests, tree cover in this small Central American country fell from 80 percent in the 1950s to just 21 percent in 1987. It was, for a while, being deforested faster than any country on the planet. For years, environmentalists argued that this was the "inevitable consequence" of a population that at one point had doubled in seventeen years—another world record. The only thing wrong with that statement was the word "inevitable."

Since 1987, Costa Rica has been regrowing its forests. Today tree cover is back above 50 percent, even though the population has grown more in the two decades since 1987 than in the two decades before. The nation now pays farmers to protect the forests rather than to chop them down, and gets extra income from millions of tourist coming to see the jungle wildlife. "We discovered that it was government policies that were destroying the forest, not too many farmers. This is true across the world," says former Costa Rican environment minister Carlos Manuel Rodriguez. That is an important lesson, and one which environmental pessimists miss. There is another way.

· · ·

The most global threat we face today is climate change. So what chance do we have of fixing that before it fixes us? The task looks daunting. Climate scientists say the world needs to cut greenhouse gas emissions by at least 80 percent by 2050 to curb dangerous climate change. That means transforming how we generate and use energy—in homes, factories, offices, public spaces, and transportation. It requires a combination of new energy technologies that do not release carbon dioxide into the atmosphere. And it requires redesigning our lives and living environments to reduce energy needs—for instance, building cities where local services can be reached on foot and the rest by mass transit systems rather than cars.

We need to spread the new ideas and technologies fast. And they need to arrive fastest in the countries currently developing their energy infrastructure most rapidly. The UN estimates that by 2030, $26 trillion will be invested in energy worldwide, and more than half of that will be in developing countries, where two billion people still lack electricity supplies. Those countries need to leapfrog to the new technologies without passing through the dirty and polluting phase that most industrializing countries have taken.

The good news is that we know most of the technologies that we require. Wind energy is well developed and not expensive. Solar power is fast emerging. There is growing interest today in concentrated solar power, which uses mirrors and lenses to focus solar energy to heat water that then runs conventional power turbines. Large areas of desert, from Nevada to Algeria to India, could be covered in mirrors catching the sun's energy. Other natural sources of energy that can be harvested include tidal and wave power and geothermal energy (hot rocks). We could keep burning fossil fuels if we found ways to capture carbon dioxide emissions and bury them out of harm's way. Nuclear power and hydroelectricity have their detractors but will live on.

Future vehicles are likely to be driven by electricity. If the electricity is generated without emitting carbon dioxide, that will be a huge gain. Biofuels have been justly criticized for taking land and water needed for growing food. But future biofuels, particularly those using algae or waste products from farming, may be a better bet. My guess is that their main use will be to power planes. Above all, there is huge potential, in almost every sphere of life, for much greater efficiency in the use of energy. From heavy industry to transportation, buildings, and consumer electronics, cost-effective modifications and redesign can cut energy use, typically by 30–50 percent. The world is switching to energy-efficient light bulbs right now. But almost every other use of energy could make a similar step change at similarly negligible cost.

There is no inevitability about more consumption requiring more energy, or about more energy resulting in more carbon emissions. The link can be, and sometimes is being, broken. An international comparison of how much wealth different countries currently create (in terms of gross domestic product) for every ton of carbon dioxide they emit is revealing. Both Russia and China produce only around $400 of GDP per ton of emissions. Five times better are the United States and Australia, which each produce around $2,000 of GDP per ton. Britain, Germany, and Italy do still better at around $3,500 per ton, while nuclear-powered France is above $5,000. Sweden manages $6,000 per ton, and two countries as different as Switzerland and Cambodia both manage to produce around $9,000 of GDP for every ton of carbon dioxide emitted. If the whole world did as well as those two countries, then the world's carbon dioxide emissions would be only a third what they are today. So cutting emissions by 80 percent by 2050 is not impossible. It is doable. Today.

. . .

We have the capability to tackle these great issues and to carve out a sustainable future. In the next chapter, I will look at whether we can continue to feed ourselves. But we have to persuade ourselves of the seriousness of the threats we face, so that we act. If necessity is the mother of invention, we have to recognize the necessity. We have to fear those "limits to growth" in order to defeat them. Just as the green revolution staved off mass global hunger, so a new revolution to change the way we generate and use energy can stave off the worst of climate change—if we have the will to do it.

For me, environmentalists are at their best when they alert us to the dangers—and at their worst when they succumb to the belief that their worst predictions are destined to come true. The optimists are at their best when they convince us that anything is possible—and at their worst when they are convinced that we don't have to change in order to achieve it, that all we need to do is trust God or the markets.

There is a paradox here, of course. Half a century ago, Vogt, Huxley, Ehrlich, and the rest were wrong to predict mass hunger in the late twentieth century. But it was their dire predictions and the world's response to them that ensured it didn't happen. The appalling thought of billions of dead galvanized a generation. The challenge now is to prove the doomsayers wrong again. If we do it right, everyone can live a good life. As Mahatma Gandhi famously put it, "There is enough for everyone's need, but not for everyone's greed."

CHAPTER 24

Feeding the World

For most of human existence, the land appeared limitless. Whenever populations grew too large for comfort, societies occupied new land. Civilizations sometimes collapsed, with a spectacular fall from grace for priesthoods and military elites, bloated bureaucracies and royal families. But whether it was the people of Angkor in Cambodia or the Maya in Central America or the Mesopotamians in the Middle East, the mass of their subjects generally shrugged their shoulders, got on their horses or donkeys, and headed for pastures new. Now that humanity is counted in the billions, the wide open spaces have been eaten away and the frontiers tamed. There is nowhere else to go.

The last great occupations of new land occurred in the hundred years between about 1860 and 1960. During that time, farmers brought under the plow the American Great Plains, the South African veldt, the Russian steppes, and the pampas of South America. They also drained marshlands, irrigated deserts, and took chainsaws to the once-forbidding jungles. Growing populations made this annexation of wilderness necessary, and advancing technology made it possible at rates Malthus never envisaged. These were the high days of "extensification"— growing more food by taking more land.

But by the 1960s, most of the best land was taken and the frontiers were being pushed up inhospitable mountainsides and onto poorer soils, including those beneath the tropical rainforests. The game seemed to be up. The battle to feed the world was lost, Paul Ehrlich famously declared. But human ingenuity stepped in again. This time, "extensification" was replaced with the "intensification" of the green revolution. We deployed technology not to tame new lands but to make far more effective use of the farmland we already had. In the past half century, the world has added just 10 percent to farmland but more than doubled food production. This was done with new varieties of crops and ever-increasing volumes of water, fertilizer, and pesticides. More than half of all the chemical fertilizer ever added to the world's soils has been added since the late 1980s.

The amount of cropland per person has shrunk globally from 0.57 acres half a century ago to 0.27 acres today. The United States still has 0.52 acres per head, but India largely feeds itself with 0.25 acres per head, and China does likewise with 0.17 acres. What next? We have fed nearly seven billion people. But are we doing it in a way that destroys the future viability of soils and water supplies? Can we go on to feed eight or nine billion?

The bad news is that the scientific crusade of the green revolution has faltered in the past decade. Grain yields grew by more than 2 percent a year between 1950 and 1990, exceeding population growth. But since then, yields have increased at little more than half that rate. The revolution became a victim of its own success. As the production of food rose faster than population, food prices slumped. Governments got complacent, and investors pulled out. In the 1980s almost a fifth of foreign aid went to agriculture. But by 2006 that figure had shrunk to 4 percent.

The research budget of the International Rice Research Institute has been halved since the mid-1990s. Once, the institute employed five entomologists helping to breed plants resistant to insect pests. Now it has one. "People felt that the world food crisis was solved, that food security was no longer an issue, and it really fell off the agenda," says Robert Zeigler, its director. At the other crucible of the green revolution, the International Maize and Wheat Improvement Center (CIMMYT) in Mexico, researcher Hans-Joachim Braun complained in 2008, "You can't even mention increased yield in a research proposal and expect to get it funded."

The fulcrum of politics and economics has moved to the cities, and farming has languished. This is true even in countries where there are food shortages. African governments spend only 5 percent of their budgets on agriculture. The rules of free global trade make such neglect almost compulsory. Rich countries insist on their right to flood the markets of poor countries with cheap food. The International Monetary Fund and World Bank have forced closure of most of the state marketing boards that once guaranteed prices for domestic farmers, and they have demanded cutbacks in subsidies for essentials like fertilizers.

The world was brought up short in 2008 by soaring food prices in international markets. Politicians were unnerved as food riots broke out in thirty-two countries, from the Ivory Coast to Mozambique, Mauritania to Bolivia, and Indonesia to Yemen. In Haiti, rioters forced the prime minister from office. "We are seeing a new face of hunger," said Josette Sheeran, head of the UN's World Food Programme. "We are seeing urban hunger, with food on the shelves but people unable to afford it."

Some blamed the sudden crisis on droughts in Australia. Others blamed the diversion of U.S. corn to make biofuels, or surging oil prices that increased the cost of fertilizers, or meat eating in China, or commodity speculators. Whatever the immediate cause, the soaring prices underlined the fragility of an increasingly global market in food—and its political sensitivity. The M-word began to reappear, with the suggestion that the world is finally reaching limits on food production.

Prospect magazine in the UK put the headline "The return of Malthus" above an article by a top British food industrialist, Chris Haskins, who said, "The Malthusian predictions were wrong for 200 years, but might prove right in the next 50 years." A commodities newsletter warned, "Malthusian catastrophe coming soon."

Such talk set off a flurry of fear in rich countries dependent on food imports. They began buying up spare farmland round the world. "The time may come when even if you have the money, acquiring some commodities [like food] will not be easy," said the director general of the Abu Dhabi Fund for Development, Mohammed al-Suwaidi, as he bought up 75,000 fertile acres of Sudan. Uzbekistan and Senegal were also on his shopping list, he said. China acquired three million acres in the Philippines and more land in Laos and Kenya. The United Arab Emirates bagged more than 1.2 million acres in Pakistan, and Saudi Arabia over a million more in Tanzania. In mid-2009 a group of South African businessmen was reportedly about to buy twenty million acres in the Democratic Republic of the Congo, and the UN reported that fifty million acres, an area half the size of Europe's arable land, had been sold or negotiated for sale to foreigners in the previous year.

Are these fears well founded? Up to a point. We are certainly undermining our ability to feed the world, through damaging our water and soils and by altering climate in ways that could damage future crops. We are using more than half of all the world's river flows each year, mostly to irrigate crops. As rivers run dry, farmers are pumping up underground water, most of which is not being replaced by the rains. A fifth of India's food is grown by "mining" this groundwater. Likewise, the water beneath the fields of northern China, which feeds half of China, is being pumped dry fast.

Equally disturbing is the loss of topsoil. Often only six inches thick, it is literally the difference between civilization and starvation. By some estimates, a third of the world's fields are losing soil faster than natural processes can create new soils from the bedrock beneath. And now comes the biggest threat of all: climate change, which will create new droughts and raise temperatures, threatening crops round the world. The U.S. National Academy of Sciences reckons that every rise of one degree worldwide will cut rice, corn, and wheat yields by 10 percent, with the biggest falls likely in the hungriest countries least able to adapt.

But we should avoid a sense that all this is inevitable. It is not. Bleak though the bald figures are, they are no worse than those in the 1960s. Just as then, they reveal not natural limits but the current limits of our competence, both political and technical. Feeding the world in the twenty-first century requires doing things dramatically better.

· · ·

In the first eight years of the new millennium, global grain production grew by 1.2 percent a year. That is less than before, but still almost exactly the same rate at which the world's population grew. The green revolution has faltered, but it is still keeping pace with population. The trouble is that consumption of grain grew in those eight years at 1.6 percent a year—substantially faster than population. The rising demand was driven by the world's growing appetite for biofuels and for meat and dairy products. By 2008 a third of the corn grown in the United States, one of the largest producers, was fueling cars and not people. As U.S. environmentalist Lester Brown points out, filling an SUV tank with biofuel takes corn that could feed one person for a year. Meat is not much better. It typically takes eight calories of grain to make one calorie of meat.

Of the two billion tons of grain grown round the world in 2008, under half was eaten directly by people. Paradoxically, this is good news, says U.S. demographer Joel Cohen. "We know we can feed 10 billion people, because we are already growing enough, if they have a vegetarian diet." But it begs big questions about whether, in order to ensure everybody gets sufficient food to eat every day, we will sacrifice meat consumption or biofuels production. It certainly underlines my central point in these chapters on limits: the real threat is consumption patterns, not overpopulation. But at least we know the world can be fed. And if we can improve yields some more, we can feed the world and still have capacity for meat and even some biofuels.

A second cause for optimism is that farm yields in most of the world are a small fraction of the potential using existing seeds. In Africa, yields are typically less than half a ton of grain per acre, in Asia one ton, and in Europe and North America two tons or more. Futurologist and arch optimist Jesse Ausubel of Rockefeller University in New York says that "if during the next fifty years or so, the world's farmers reached the average yield of today's U.S. corn grower, ten billion could be fed with only half of today's cropland, while they eat today's U.S. calories."

That requires good soil, fertilizer, and water, of course. But there is good news here too. The flip side of our reckless management of water and soils is that we could do things so much better. Conventional methods of flooding fields to irrigate crops waste most of the water. But if they were replaced by simple drip irrigation systems—which deliver drops of water down pipes close to crop roots—most of the world's farmers could halve their water use. As we come up against real, life-threatening limits to water supply, we need such innovations. It's not rocket science. It's just tubes with holes in them. But if we don't do it, the results could be billions of hungry people.

· · ·

Globally, then, there is cause for some optimism about the world's ability to carry on producing enough food. There is no inevitable Malthusian disaster ahead. But there are plenty of possible disasters that are not inevitable—if we get things wrong.

That said, it is one thing to ensure there is enough food on the global dinner table, but quite another to make sure everyone has a seat at the table. This applies especially to people who lack the money to buy food and who are dependent on food grown by themselves or their neighbors. These people make up the majority of the world's hungry. They are also those most dependent on the health of their local environments. For them, the global market is often a fantasy or a destructive interloper. It matters little to them whether the global grain warehouses are full if their village granaries are empty. They rely on the local environment to deliver. If their soils and water supplies are overused or trashed, the result will be hunger, poverty, famine, and sometimes mass migration. So regardless of whether the planet itself can produce enough food, these people need to be able to sustain themselves.

The green revolution was designed to maximize global food output. The next revolution needs to get local. It needs to help these poor farming communities, the ones largely left out of the last green revolution, to find ways to manage their own soils better—using livestock to fertilize soils, conserving rainwater on their land in case of drought, breeding and exchanging local crop varieties, and finding natural predators for troublesome pests.

In particular we are talking here about Africa, the place about which some of the most apocalyptic demographic warnings are made. Africans cannot feed themselves now, yet their populations are still soaring. How can twice as many Africans feed themselves on the same continent that cannot sustain today's population? It's a good question. So let's take a closer look at the Dark Continent.

Exhibit A in the case against African survival is the small and densely populated central African state of Rwanda. There, in 1994, up to a million cattle-herding Tutsi people were murdered over about a hundred days by militias of the country's Hutu farming majority. Most victims were unarmed and hacked to pieces with machetes. It is often called the first modern Malthusian disaster. The argument is that population pressure cooked up the genocide. James Gasana, who was the country's minister for agriculture and the environment just before the slaughter, has asserted that rapid population growth created an environmental crisis that "produced" the conflict. But is it true?

Rwanda's unimaginably brutal bloodletting was the culmination of a civil war that had been rumbling on for four years. After Rwanda's president, Juvénal Habyarimana, died in a plane crash that was presumed to be an assassination, tensions exploded. In his book *Collapse*, Jared Diamond of the University of California at Los Angeles goes straight to the point in a chapter he calls "Malthus in Africa."

Rwanda was the most densely populated country in Africa. Its population had quadrupled since 1948. Fertility had been extremely high at 8.5 children per woman in 1983, though by the eve of the massacre it had fallen to 6.2. Production of food had grown as farmland had been extended up valley slopes. Nobody was starving. In fact, the amount of food grown per capita had been increasing. But "by 1985, all the arable land outside national parks was being cultivated," Diamond writes. Limits had been reached. Young people, finding it difficult to get land of their own, delayed their marriages, creating "lethal family tensions." There was a "Malthusian dilemma: more food, but also more people, hence no improvement in food per person. . . . Modern Rwanda illustrates a case where Malthus's worst-case scenario does seem to have been right."

French historian Gerard Prunier has concluded more cautiously that "the decision to kill was made by politicians. But at least part of the reason why it was carried out so thoroughly by the ordinary peasants . . . was the feeling that there were too many people on too little land, and with a few less there would be more for the survivors."

It sounds convincing. But we need to be careful here. Access to land was certainly part of the struggle, but the problem for the Malthusian analysis is that it wasn't the landless who took up arms. Gasana, the former minister, says that "wealthy northern Hutus and their allies spent much of the 1970s and 1980s accumulating land for their own estates . . . reducing the amount of land available for [Tutsi] peasant farmers." So it was the people who committed the genocide who had the land. The people who had been losing it were the victims, not the perpetrators.

Rising population certainly aggravated tensions. But the simplistic argument that overpopulation created land shortages that created food shortages that created genocide just does not stand up. Though Rwanda is a very densely populated country, this is largely because its volcanic soils are also very fertile and rains plentiful. It can sustain a large population. Through the 1970s and 1980s, Rwanda was one of sub-Saharan Africa's top agricultural performers. Food production rose by 4.7 percent a year for almost two decades, easily outpacing population growth. On the eve of the massacres, Rwanda was one of the best-fed countries in central Africa, averaging more than two thousand calories a day per person.

If hunger wasn't the trigger, the country wasn't succumbing to environmental degradation either. Farmers living in the areas with the highest population density had been improving their farming methods by conserving soils and planting trees. The year before the genocide, Robert Ford, a development scientist then working in the country, wrote a paper describing a "positive association" between population growth and the extent of woodlots in Rwanda.

Other things besides land were making rural Rwandans poor. The ongoing war had caused massive migrations of people escaping the fighting. And there was coffee. It had accounted for three-quarters of Rwanda's export earnings and was mostly grown on small plots. In 1989 the world price for coffee dropped by half as price controls were swept away in the name of free markets. I have never heard the free market in coffee mentioned as an explanation for the Rwandan genocide, but arguably it was much more important than growing population.

Thomas Homer-Dixon, an analyst of environmentally driven conflict at the University of Waterloo in Canada, concluded that in Rwanda such theories are "too simplistic. . . . The environment and population had at most a limited aggravating role." I agree.

·   ·   ·

The case of Rwanda is important because it has become a totem for pessimists about Africa's ability to support its growing population. One old Africa hand, John Guillebaud, a former chairman of Britain's Optimum Population Trust, has argued that "for the whole planet to avoid the fate of Rwanda, Malthusian thinking needs rehabilitation." Another British public health professional with bags of experience in Africa, Maurice King of Leeds University, argues that "large parts of sub-Saharan Africa are demographically trapped . . . committed to a future of starvation and slaughter." Rwanda is his model. When pressed, he also names Niger, Ethiopia, Malawi, Uganda, Burundi, and eastern Congo. He says these countries are trapped, because they have too many people to prosper without a reduction in numbers. Guillebaud, a family planning doctor, says that for such countries, "once environmental carrying capacity is grossly exceeded, the only likely outcomes are starvation, disease, interethnic violence and genocide, migration, and/or dependence on aid from the international community."

I think this pessimism is dangerous, for two reasons. For one thing, it echoes the Malthusian fatalism that the British used to excuse their inaction during the Irish potato famine a century and a half ago: "nothing to be done . . . too many people . . . brought it on themselves . . . better let the carnage play out." It almost suggests that something like the Rwandan genocide might be a good thing in the long run.

More importantly, the idea of overpopulated Africa simply is not true. The continent contains eleven of the world's twenty least densely populated nations and only one of the twenty most densely populated. That last is Mauritius, which is also one of Africa's richest countries. Africa's problem is bad agriculture, not too many people. Yes, rising population is sometimes exposing how bad its agriculture is. But no, the continent is not trapped by Malthusian limits to what it can do to feed itself.

When Africans start to get things right, the results are often spectacular. In 2005 Malawi suffered a dreadful drought that left millions hungry. In response, the government provided its farmers with better seeds and more fertilizer. Nothing too fancy, but it was effective. The result was that Malawi doubled its harvest in one year. Jeffrey Sachs of Columbia University says just ten billion dollars could do the same for the whole of Africa.

Bob Watson, chair of the International Assessment of Agricultural Knowledge, Science and Technology for Development, which reported its findings in 2008, also refutes any Malthusian limits to feeding Africa. "Today's hunger can be addressed with today's technology. It's not a technical challenge, it's a rural development challenge." He says farm yields across the continent can be raised from less than half a ton per acre to as much as two tons.

In fact, good news is not hard to find in Africa—and often in places that were in the past written off. The big surprise is that many parts of Africa could gain from having more people, not fewer. This is counterintuitive, I know. But for every potential Rwanda in Africa, there is a potential Machakos.

Machakos is a rural district of Kenya a couple hours' drive east of Nairobi. Sixty years ago, British colonial scientists condemned its treeless hillsides of eroding sandy soils as an ecological basket case. The hills were "an appalling example" of environmental degradation, soil inspector Colin Maher wrote in the 1930s. He blamed this on the "multiplication of the natives." The Akamba had exceeded the carrying capacity of their land and were "rapidly drifting to a state of hopeless and miserable poverty and their land to a parched desert of rocks, stones and sand." Other colonial administrators produced similar reports at regular intervals until Kenya won independence in 1963. Machakos seemed doomed by Malthusian forces.

Well, look at it now. Since independence, farm output in Machakos has risen tenfold, yet there are also more trees and soil erosion is much reduced. There are bad years when the rains fail. But the Akamba, still using simple farming techniques on their small family plots, are today producing so much food that when I visited, they were selling vegetables and milk in Nairobi, mangoes and oranges to the Middle East, avocadoes to France, and green beans to Britain. For them, multiplication of their numbers has been part of the solution rather than the problem. They have sprung the demographic trap.

We know about Machakos because British geographer Michael Mortimore and development economist Mary Tiffen have studied the area in detail. They wrote a book called *More People, Less Erosion*, which concluded that in Machakos it was a growing population that had delivered the manpower, or more often womanpower, to improve the land. There may have been more mouths to feed, but there were also more hands to work. Those hands dug terraces on hillsides to reduce soil erosion, created simple structures like sand dams and field embankments to capture rainwater, planted trees, raised animals that provided manure for the fields, and introduced more labor-intensive and high-value crops, such as vegetables. It was a homemade green revolution of just the kind that Africa needs.

Mortimore believes that the true demographic trap in much of Africa is underpopulation. More people, he says, can deliver more Machakos miracles built on "specialisation, diversification of the economy, rising living standards and an increasing rate of technological change which [can] outpace any threat to the depletion of resources." That might be pushing the evidence a bit far, but it is clear that African farmers can use the extra manpower available in larger populations to manage their land better and feed more people. Human ingenuity, "impelled by population growth and made feasible by additional labour," can outwit environmental decay, says Mortimore. Ester Boserup would say: I told you so.

When I first asked mainstream development economists and NGOs about Machakos, I was repeatedly told that it was a one-off—that across Africa, the rising numbers of Africans are stripping the land bare of wood, overgrazing its pastures, and plowing the soils to death. Sure, such things happen. But in most places there is a very different narrative. Far from stripping their land bare of trees, Kenyan farmers are planting on such a scale that one study found "planted woody biomass" has increased by 30 percent. And the planting is greatest in the most densely populated areas.

In the highlands of western Kenya, the Luo people are packed in at densities approaching those in Rwanda. But the farmers showed me how they were replacing their fields of maize with a landscape richer both commercially and ecologically. They had planted woodlands that produced timber, honey, and medicinal trees. I saw napier grass, once regarded as a roadside weed, sold as feed for cattle kept to provide milk.

Maybe Kenya is special. After all, it is home to environmentalist Wangari Maathai, who won the Nobel Prize for her efforts to encourage tree planting. But the evidence suggests not. Two young British researchers, Melissa Leach and James Fairhead, found that across West Africa, there are today more forested areas than half a century ago—because farmers are planting trees.

For the past thirty years, the fringes of the Sahara desert have been in the cockpit of environmental doomsaying, with fragile soils, unpredictable rains, and fast population growth. When the rains have failed here, there have been famines. Countries like Mauritania, Niger, Ethiopia, Mali, and Sudan remain desperately poor and have some of the world's fastest-growing populations. Niger and Mali have the two highest fertility rates in the world.

Nobody can doubt the damage that can be done by growing populations in such places. But it isn't always so. Chris Reij, a Dutch geographer at the Free University in Amsterdam, has worked for years in the desert fringes of West Africa, sometimes with Oxfam. He too sees ecological recovery amid rising populations. "Back in the 1980s and early 1990s, a lot of the land was treeless. There had been frequent droughts. Farmers had chopped down their trees for firewood, and the desert was spreading. But when I went back in 2004 and again in 2006, I drove for eight hundred kilometers east from the capital, Niamey, and there were trees everywhere. I reckon Niger has gained two hundred million trees in the last decade."

Villagers saved the situation by ignoring advice from European agricultural experts, who had told them to remove all the trees from their fields so they could grow more crops. They figured removing trees was killing their land. They were right. Reij says, "Within a few years, the trees were growing high enough to protect the crops against the winds and to help stop the sands from spreading. The trees now also provide fodder for livestock, so farmers get more manure for their fields."

Tending trees and shoveling muck take work, of course. They take people. But in the 2005 drought in Niger, the trees provided both food for livestock that would otherwise have died and cash from the sale of fuel wood. Just as in Machakos, more people are turning out to be good for the land. "The idea that population pressure inevitably leads to increased land degradation is a much-repeated myth. It does not. Innovation is common in regions where there is high population pressure. This is not surprising. Farmers have to adapt to survive," says Reij.

· · ·

There will be exceptions, distressing situations where farmers are unable to rescue their declining environments. There will be places where fast-rising populations trigger a dangerous tailspin of decline, and where land disputes, war, and bad government leave communities incapable of harnessing their human resources. There will be environmental refugees: victims of spreading deserts and deforestation, lack of water, disappearing fisheries, floods, toxic spills, hydroelectric projects, strip-mining, soil erosion, and much else. Norman Myers of Oxford University, an environmental academic who has popularized the concept of environmental refugees, claims that such refugees are already numbered in the tens of millions. "Each day roughly five thousand people find themselves obliged to abandon their homelands for environmental reasons," he says. Climate change may force 200 million to move, he has suggested.

I have met people in India moving because they had emptied the underground water reserves in their villages, Bangladeshis who went to the city because their farms were engulfed by rising tides, people who left the shores of the Aral Sea because the irrigation water ran out, and Filipino fishing families who moved to other islands when their coral was wrecked. They left because their land could no longer feed them.

Environmental refugees till now have mostly moved locally, within their own countries. But the real fear among westerners is that the refugees may go international. The first such modern "flood" may have come out of Haiti in the 1990s, when tens of thousands of Haitians jumped into boats and headed for Florida. There was an outcry. Florida had welcomed Cuba's refugees as escapees of communism. But Haitians were seen as environmental wreckers who had destroyed their country by cutting down all its forests and destroying its soils. American anthropologist Catherine Maternowska in 1994 called Haiti "an island of environmental refugees."

Desecrated first by French colonialists, who stripped the island of timber and filled it with slaves to work sugar plantations, Haiti is today the poorest country in the Western Hemisphere, and its fertility rate, at almost five children per woman, is the highest in the Americas. More than a million Haitians have left since the 1950s. Many more live in the degraded, crime-ridden slums of Port-au-Prince and subsist on the half billion dollars in remittances sent home each year by relatives working in the United States. Foreigners refuse to invest. Tourists don't come.

Virtually all the land that can be farmed is under the plow, parceled into tiny family plots and receiving few inputs. Haiti's forests are gone. Many of the fields are on steep slopes, and the soils get washed into the valleys in the rains. Parts of the country are almost literally turning to desert. Farm yields are so low that most of the food has to be imported. Poor government, poverty, and environmental decay are combining in a way steeped in foreboding for other nations. It is a frightening advertisement for what can go wrong. I don't deny it. But neither Africa nor Haiti are stuck in a Malthusian trap. And to suggest they are doomed is a lie in many ways as demeaning and wrong as Malthus's assertions about the English working class two hundred years ago—and as potentially dangerous as the way Britain washed its hands of the Irish famine.

Demography may help drive communities to crisis, but it does not define how they respond.

CHAPTER 25

Slumdogs Arise

I sat in the lobby of the Oberoi Hotel in downtown Mumbai in the summer of 2008. The hotel is a symbol of the thrusting, modern, would-be tiger economy of India. With me was Perianayagam Arokiasamy from the Mumbai-based International Institute for Population Sciences. Around us in Mumbai, some twenty million people were living in cardboard cities down by the shore, swish high-rise apartments, neat middle-class suburbs, some of India's largest slums, and the world's first two-billion-dollar house. They coursed through the city on buses and trains and rickshaws, in taxis and cars and on foot.

Mumbai, a city of twenty million people, is the beating heart of boisterous, cacophonous, industrializing, investing, plotting India. A youthful city; a fast-growing city. As we chatted, somewhere across town they were completing *Slumdog Millionaire*, the Oscar-winning rags-to-riches movie about a slum boy from the city who wins the game show *Who Wants to Be a Millionaire?* What better symbol for the new optimism of modern India? But as we spoke, somewhere out there too (or possibly in one of the hushed air-conditioned rooms above us), some young Pakistanis were plotting a November surprise. A few weeks later, they would rampage through the business district, mowing down commuters at the main railway station and backpackers in an Internet café before laying siege to its two most prestigious hotels. The rampage killed 120 people, including many in the Oberoi itself.

· · ·

Cities are where population growth shows itself. Especially mega-cities, defined as cities with ten million people or more. The world didn't get its first megacity until sixty years ago, when New York City passed the landmark figure. Today there are at least twenty of them, of which Mumbai is among the top five. Cities are now home to half the world's population. They are also the economic powerhouses of the world. They consume three-quarters of the resources we take from nature and produce a similar proportion of our pollution. They sound like eco-pariahs, and they are. But they are part of the solution too.

For the demographic growth of cities contains a paradox. Cities have low fertility. Globally, rural fertility is about 3.0 children per woman, while urban fertility is about 2.2, roughly the replacement level. But it is rural populations that are stable, while those of cities are growing fast. This paradox arises partly because cities are engulfing formerly rural areas, but mainly because tens of millions of people are moving from the countryside to the cities each year. Thanks to these migrants, there are a hundred more urbanites on the planet every minute, most of them young, streaming from bus stations and railway platforms, arriving in cars and vans and on motorbikes and by foot, often to live in slums and shanties, but always with the hope of something better.

Like Dick Whittington, the character in the British rags-to-riches folk tale who seeks his fortune in medieval London, those with ambition head for the city. The streets may not be paved with gold, as Dick hoped, but at least they are paved. Urban dwellers have higher incomes and by most measures live healthier, easier lives than their rural counterparts. Urban children have a 25 percent better chance of survival to adulthood than those left behind in the villages.

Cities have a large ecological footprint. They call on resources over a wide area to provide food and raw materials. London's footprint is 120 times the size of the city, drawing on resources from the factories of China, the wheat prairies of Kansas, the tea gardens of Assam, and the copper mines of Zambia, among other places. Locally, cities put huge strains on natural ecosystems, polluting rivers and coastal waters, consuming forests and water, degrading soils, disrupting drainage, and stunting crops. Urban smog and acid deposition in China are estimated to be reducing crop yields in the countryside by up to a third.

Cities—already the largest, most complex man-made structures on the planet—are growing into wider urban zones taking over ever-larger areas of the planet. Single megacities are being replaced by what some call "urban archipelagos" and the French geographer Jean Gottmann dubbed "megalopolises." Mexico City is engulfing surrounding cities like Toluca and Cuernavaca. Kolkata is dispersing across West Bengal. London has spawned an urban commuter region across southeast England that stretches west toward Reading and Oxford, north toward Cambridge, and east along the Thames estuary. São Paulo is embracing a new "golden urban triangle" stretching to Rio de Janeiro and Belo Horizonte.

Tokyo is extending out to Japan's second megacity, Osaka, creating a megalopolis of seventy million people linked by bullet train. Shanghai is reaching across the Yangtze delta to Suzhou, Nanjing, and the hundred miles to Hangzhou, which will soon be just twenty-seven minutes away on a new maglev train. In southern China, Shenzhen is already contiguous with Dongguan and Guangzhou and is poised to join Hong Kong and Macau in a huge megalopolis spanning the Peark River delta. Indonesian planners foresee Jakarta stretching for six hundred miles across the north size of Java, capturing 120 million people.

There is no sign of urbanization slowing. Demographer Joel Cohen predicts that the world's urban population will double from today's three billion to six billion by 2050, the equivalent of adding a city of a million people—the size of Birmingham in England or Detroit in the United States—to the planet every week. The urban world is not only growing at an extraordinary pace, it is also absorbing the youths of the planet. Cities embody the hopes and fears of that generation, perhaps of us all.

Two critical questions for cities are whether the wealth they generate can justify their large ecological footprint and whether we can reduce that footprint. A well-run urban sector can ensure national prosperity; a badly run sector can become a drag on the whole country. And cities do have potential green attributes. Well-planned cities can take advantage of their high population densities to share resources and reduce energy consumption—by collecting trash for recycling and developing mass transit systems, for instance. Many cities include large areas of often highly productive agricultural land amid the highways and high-rises. Up to a fifth of the world's food is grown in "urban" areas.

Cities are also the great innovators, the great investors, and the great drivers of change. That is partly because innovators go there to get ideas and feed off each other. Far from being polluting parasites, cities could hold the key to sustainable living: green innovation and low fertility.

. . .

Some fear the slums. They can be dangerous places. The biggest causes of death among young people in São Paulo are traffic accidents and homicide. A Californian urban geographer, Mike Davis, has written a book called *Planet of Slums*, an apocalyptic take on the huge slums that dominate many megacities in the developing world. It is terrifying. But his image is not what I see when I go to slums. They have their gangs and drugs and open sewers and heartbreaking stories. But slums are at least as much places of hope and enterprise and innovation. That's why people move to them. For every gun-toting gangster or terrorist, there are a hundred romantic would-be slumdog millionaires. Hormones aren't all bad. Even male hormones.

I went to Dharavi in Mumbai, where the movie's fictional slumdog was brought up. Often called Asia's largest slum (in fact, that dubious distinction goes to Orangi in Karachi), it has 600,000 people crammed into a maze of narrow lanes and shacks covering less than one square mile, about half the size of New York's Central Park. It is "a vision of urban hell," according to *Smithsonian* magazine. Visiting businessmen shiver at the thought that terrorists hiding in Dharavi could shoot down a plane taking off from Mumbai International Airport right over the back fence. The municipal government wants to bulldoze it and start again. As do real estate developers, who lust to replace it with a posh estate of high-rise apartment blocks, like the one just over the river. The inhabitants? They want to stay, because Dharavi is a thriving community, entirely unlike the terrifying image.

It is cramped and higgledy-piggledy, but surprisingly ordered. Most of the houses have two stories, painted exteriors, and sometimes paths to the door. Down one alley, Mildred was peering down the lane from her front door as I walked past. Inside her house were two tidy rooms, one with a sink and tap. She had tenants living upstairs. It was crowded, but "this area is like a family," she said. "I know if I leave the house, the neighbors will look out for my girls." Her two daughters are thirteen and seven years old.

No, she doesn't fear crime or violence any more than she would anywhere else. Mildred was born in Dharavi, but her father, who worked for Indian Airlines, came here from Tamil Nadu. Her husband was in the navy. Her eldest daughter planned to be a teacher. "She has a crush on her own teacher," Mildred told me with a smile. They could be in any city anywhere in the world.

With three sides of the slum bounded by suburban railways and the fourth by a main road, getting round the city from Dharavi is easy. But most people living in Dharavi work here too. There are workshops down every lane—by one count ten thousand of them, mostly single rooms above the houses. I found dozens of people baking and selling rice cakes; a small colony from Rajasthan firing ceramic pots; Muslims from Uttar Pradesh finishing leather handbags; textile machinery rattled away in upstairs rooms, stitching decorative patches onto denim jeans; and a man screen-printing logos onto shirts for brands like Hugo Boss and Honda.

Down by the canal was a warren of tall corrugated-iron sheds, where hundreds of workers were sorting and processing Mumbai's waste. The discharges from their workshops into the canal were foul, but there was lots of work and a relaxed atmosphere. I saw workers chopping plastic bottles into pellets, smelting aluminum cans into small ingots, turning heavy cardboard boxes inside out and restapling them, washing cooking oil cans, recycling bits of old hotel soap into new bars, and sifting through general rubbish to extract ballpoint pens, plastic pots, metal jar tops, and different sorts of plastic.

Dharavi may be a slum, but it performs vital activities for the city and has more services for its residents than an average European housing estate. It has schools and clinics, drug dispensaries and bakeries, hairdressers offering any styles you want, and all manner of food and hardware stores. Early in the afternoon, the children all came home from school, walking through the lanes in neat uniforms with satchels on their backs. Urban hell? I couldn't imagine that the *Smithsonian* magazine writer had actually been here.

Most homes have power and water. I met R.J. Shanmuganana, who, like Mildred's father and many others, came to live here from rural Tamil Nadu. "My father sent me because I wasn't interested in the village. If I'd stayed in the village, we would have had to divide the land among five brothers, so it is good that I went," he said. R.J. settled down in Dharavi and became a security guard at Mumbai's nearby atomic research academy. Recently he joined the staff of a housing co-op, one of eighty-nine similar organizations in Dharavi. He was, he said, engaged in the constant task of improving the slum housing stock. Self-help was the name of the game here.

They do tours for foreigners through Dharavi. On the edge, close to the road, they sell leather goods and jewelry to tourists, who pay with credit cards. Middle-class Indians often complain about this "poverty tourism." One woman told me, "It gives the wrong idea about our city." But she had never been there, and I think those living in the better suburbs have bought the propaganda and would just rather Dharavi wasn't there. Anyhow, after *Slumdog Millionaire*, the tourist business is booming.

There were nasty aspects, of course. The main one is that Dharavi still lacks indoor toilets. For some reason, sanitation has always had a low priority in India. And there are drainage problems in Dharavi that mean foul water can flood the lanes during the monsoon season. This is hardly surprising. When the first settlers came here sixty years ago, nobody else wanted the mangrove swamp down by a creek in the heart of the city. But these are reasons to improve the place, not to tear it down.

Back in London after my visit, there was media merriment when Prince Charles declared at a conference that the "Slumdog shanty town" was a model of green development. It was generally agreed that Charles was off with the fairies again. Well, the contrast between Dharavi and his Highgrove estate in Gloucestershire is disconcerting. But actually, he is right. I don't know if he has been to Dharavi, but his description of a minicity constructed with local materials, of walkable neighborhoods, and even of "an underlying intuitive grammar of design that is totally absent from the faceless slab blocks that are still being built round the world to warehouse the poor" was spot on.

·   ·   ·

I don't want to romanticize slums or the cities of which they are a part. The contrast between the worst settlements and the rich neighborhoods close by is often obscene. In Nairobi, the green lawns and verandas of white expat colonies like Karen mock the teeming lanes of Kibera, an area the size of London's Regent's Park that is home to almost a million people (nobody's counting; it's not even on the city maps). There are no sewers and no drains, no refuse collection and no public water supply. The common pit latrines (about one for every three hundred citizens) are overflowing because the alleys are too narrow for trucks to enter and empty them. Sanitation is normally the "flying toilet" of feces wrapped in a plastic bag being flung onto neighboring roofs. Cholera is a frequent visitor. Infant mortality, 254 per thousand, is seventeen times higher than in Karen. But bad as it is, people still want to live in Kibera. In many ways, it is better than parts of rural Kenya. Its crime is partly to be more visible.

I have been to some of the world's more notorious slums. And most are the same: poor and sometimes unsanitary, but with a great vitality and sense of purpose. I have walked with an equipment-laden photographer through a favela in Rio, wandered the lanes of Orangi in Karachi (though I wouldn't repeat that since the kidnapping and murder of American journalist Daniel Pearl after 9/11), and taken similar journeys alone in the slums of Shanghai and Istanbul, Delhi and Nairobi. Perhaps I was occasionally in danger. I never felt it.

These areas have huge reserves of knowledge, expertise, and community spirit that should be tapped rather than bulldozed. The danger of books like *Planet of Slums* is that they end up demonizing the inhabitants and aiding the cause of municipal planners and real estate developers who would like to demolish the slums. Davis has a chapter titled "Illusions in Self-Help." His argument is that unless the forces of global finance are dismantled, such slums will not be able to help themselves. That is wrong. The solidarity still found in close-knit communities like Dharavi is the best safeguard the poor have against the rich and malevolent. It is human capital of immense value. That "intuitive grammar" is not just about architecture.

PART SEVEN

·

OLDER, WISER, GREENER

The final outcome of the peoplequake will be the age of the old. The baby boomers will be senior citizens, and low fertility will ensure that the world will stay old even after they die. A majority of the world's population may soon be over fifty years old. The impacts of this longevity revolution may extinguish all others in their significance for mankind. Should we fear the rise of the wrinklies? After all, who will look after them? Or will their mature influence bring order to our chaotic world? After the adolescent ferment of the twentieth century, perhaps humanity needs the calmer, wiser influence of an aging society. It could be the salvation of both Homo sapiens and our planet.

CHAPTER 26

The Age of the Old

Ushi Okushima is the oldest resident in Ogimi, the community with the oldest population in Japan, the country with the oldest population in the world. Aged one hundred and five years in 2008, she still dabbed a little French perfume behind her ears and sipped at the local firewater before taking to the village dance floor to demonstrate traditional Japanese folk dances. Ushi was born into a country only recently departed from the era of shogun warlords. Today she works twice a week at a local store, weighing fruits and vegetables for customers scarcely younger than herself. And she holds court as the world's media come to Ogimi to discover the secrets of "longevity village." It's not a bad advert for aging. For the future of the world.

Ogimi is on the north shore of Okinawa Island, the world's longevity hot spot. It has more centenarians than anywhere else. One-third of Ogimi's 3,500 people are over sixty-five, and eleven are centenarians. Most are women, living largely independent lives. Next door to Ushi lives Matsu Taira, who just turned a hundred and still grows and cooks the vegetables that are her main diet. Is that their secret? Frugality may also help. Official statistics also have Ogimi as the poorest community in the poorest prefecture in Japan. Whatever it is, if we could bottle it, we would.

The Land of the Rising Sun has become the land of the setting sun with staggering speed. In 2005 it became the country with the oldest population in the world, with an average age of forty-three years, double what it was in 1950. Japan almost certainly has the oldest population that ever existed. One in ten Japanese is over seventy-five years old. Men can expect to live to seventy-nine and women to eighty-six. And they are healthy into old age too, taking less medication than residents of almost any other rich nation. Meanwhile, falling fertility is starting to diminish the workforce. In the past decade, the number of Japanese workers under the age of thirty has fallen by a quarter. By midcentury, Japan will probably have more people over eighty than under fifteen—as well as half a million centenarians, 90 percent of them women. Japan is leading the world into the age of the old.

A longevity revolution is upon us. Hooray? Joseph Chamie, the UN's former chief demographer, calls it "humanity's greatest achievement." A century ago, average life expectancy round the world was thirty years. In 1950 it was forty-six, and today it is sixty-six. That's an increase of between three and four years every decade. The majority of people who were alive in 1960 are now dead, but the majority of those alive now will still be alive in 2060.

The revolution is global. And despite considerable worldwide health disparities, it has been a remarkably democratic revolution. No fewer than 140 nations now have life expectancies above seventy years. Nineteen countries manage eighty years or above. A century ago, Britons could expect to live to age forty-seven. Today only eleven countries (all African, except Afghanistan) offer less than that. In the United States today you can expect to live to be seventy-eight. But you can expect to live just as long in Albania, Taiwan, and Costa Rica, and only a year less in Sri Lanka and Venezuela and Mexico and among the Palestinians on the West Bank.

Life expectancy figures give the superficial impression that the big gains in longevity are about lengthening the lives of the old. But the longevity revolution is at least as much about saving the lives of the very young. For most of human history, the average life span was thirty years or less. But even back in the days when we were hunters and gatherers, those people who made it to adulthood could reasonably expect to live to around sixty. The big problem has always been getting babies to adulthood. Losing half of all children was typical. As recently as fifty years ago, most of the developing world lost about one in seven babies before their first birthday. Today, outside sub-Saharan Africa, the proportion is almost universally below one in forty—a rate that even Britain and the United States only achieved in the 1950s.

The longevity revolution is changing the face of societies. There are more wrinkles than pimples, more walkers than training wheels, more slippers and pipes than booties and pacifiers, and more gray power than student power. A Commission on Global Aging, made up of economists, business people, and politicians, said in 1999 that the graying of society promises to "restructure the economy, reshape the family, redefine politics and even rearrange the geopolitical order of the next century." And unlike most predictions, this one is a sure-fire certainty. Barring some global pandemic, the demographics are locked in. The people that will make this happen are already here, already working for their retirement. Aging is about the only unstoppable force around.

Aging affects every country, every community, and almost every household. And it is changing geography. In southern Europe, many rural areas have become enclaves of the old. In Japan, entire villages are being abandoned because the young have left and the old are dying off. More than half of the country is now listed as "depopulated marginal land" containing *genkai shuraku,* or marginal villages, where more than half of the population is over sixty-five years of age.

Half a century ago Ogama, a once-thriving community at the end of a lane on the western shore of Honshu island, had 250 inhabitants. Now it has eight: three couples and two women living alone. All of them are retired. With no incomers, they have come up with a drastic solution. They are selling the village—its houses and rice paddies and lane and wooded hillside—to an industrial waste company. Ogama is to become a landfill. Only the Shinto shrine on the hill will survive. The cash from the sale will pay for the last residents, and their rather larger number of family graves, to move elsewhere.

Should we fear this gray wave? Can we cope with the new regiments of the aged? Or should we look forward to them? It seems perverse to fear a youth bulge on the one hand, and then to fear a preponderance of the elderly too. Let's look first at the case against the old.

The French demographer Alfred Sauvy warned some years ago that Europe could "become one enormous old people's home." Ben Wattenberg of the American Enterprise Institute, in his book *Fewer*, warned that aging is "the real population bomb" of the twenty-first century. In the United States, between now and 2030 some eighty million baby boomers will retire. By then, this silver generation will make up one in five of all Americans. Who will pay for their retirement? In 1945, the year of the first baby boomers, each retired person in the United States was supported by forty-two workers. Now that figure is down to three, and will be just two by the time the boomers are all past retirement.

In Germany, France, and Japan, there are already barely two taxpaying workers to support each retired pensioner; in Italy there are fewer than 1.3. Pension funds were becoming overstretched even before the 2008 credit crunch. The average bill for public pensions and health care for the elderly is approaching 20 percent of GDP in France and Germany. By 2030, Italy will be above 30 percent. This cannot fail to be a drag on economies.

Some economists say Japan is already suffering the economic consequences of aging. They say that the demographic window of the mid-twentieth century has turned into a demographic brick wall in the twenty-first century as Japan's growing proportion of old people drain the economy. The country's top demographer, Naohiro Ogawa of Nihon University, warned of this as long ago as 1984. I remember quoting him in an article I wrote then. "Owing to a decrease in the growth rate of the labor force . . . Japan's economy is likely to slow down, approaching an annual rate of one or even zero percent, in the first quarter of the next century." In fact, Japan suffered a financial collapse and a decade of economic stagnation in the 1990s. It has been in the doldrums ever since. No doubt other factors were involved, but the coincidence is striking.

Does the rest of the world face the same fate? The global economy as a whole has been reaping the benefit of a demographic window in recent decades. As the mid-twentieth-century baby boomers entered the global labor force, and falling fertility began to rein in the number of dependents, the world's economy prospered. But now is probably the last time in history when we will have more young humans than old humans. Starting in about 2015, we will pass a tipping point as the boomers head off into retirement. The fear must be that the baby bust will deliver an economic bust too.

Back in 2001, a conference organized by the Commission on Global Aging in Tokyo warned that the world faced "aging recessions" that could "destabilize global prosperity." Some, watching the development of the global recession in 2009, may wonder if we haven't got there.

The process of aging is furthest advanced in Europe and Japan. But the gray wave is advancing on the rest of Asia at breakneck speed.

By 2030, a third of Japan's population will be older than sixty-five, and that same age bracket in Taiwan, South Korea, and Singapore will be approaching a quarter of those countries' total population. In Latin America and the Caribbean, the proportion of old people will double by 2030, tripling in Colombia.

China already has more old people than any other country on earth, says Xiao Caiwei, international director of the China National Committee on Ageing. But the aftershocks of the one-child policy will give aging a specially unnerving trajectory there. It may turn out to have created a long-term demographic nightmare. By 2050, China will have 400 million people over sixty and 150 million over the age of seventy-five. China is used to being run by a gerontocracy, but will the oldies be able to run the farms and factories as well?

China has largely dismantled its welfare state in recent years. Like other Asian countries, it expects traditional families to look after the old. But those families have been one of the casualties of the demographic changes. Sons no longer stick with their aging parents until their death, as they once did. The whole Chinese family structure, which has survived for thousands of years, is dissolving, says Nicholas Eberstadt of the American Enterprise Institute. The little emperors, the golden generation of the one-child policy, face the prospect of having two parents each to care for, and sometimes four grandparents as well. They will have no brothers or sisters to help them, and few other relatives of any sort. Eberstadt calls it "a slow-motion humanitarian tragedy. . . . Just two decades from now, roughly a third of Chinese women entering their sixties will have no living son." The traditional support in old age will be gone at a time when the Chinese are living almost twice as long as any previous generation.

China is not alone. One night in rural Bangladesh, I had been talking to a group of poor, landless women. They seemed to me to be right at the bottom of the social heap—until they decided to introduce me to someone in an even worse state. We walked down a dark lane to a tiny shack, with mosquitoes flying noisily through the broken screen. In the shack there was just room for a bed. And on the bed, in filthy sheets, was an old woman. She was helpless. She could barely turn over and could talk only in a whisper. She was dying. She had no family and no caregivers other than the women, who fussed round her ineffectually. They spoke of getting medicines. But they had no money and no idea what good the medicines might do.

"She has no family, so we are her children," they said, adding later that they tried to take care of ten other "senior citizens" living alone locally. I was reminded of a remark by demographer Jack Caldwell when we met at his home in Canberra. Sometimes he wondered whether the fertility rate in Bangladesh wasn't too low now. "These people are still very poor. You still need kids to carry water for you and look after you in your old age." Here was the starkest possible evidence of that. I imagined this pathetic scene repeated tens of millions of times in Bangladesh and across Asia within a few decades, and hundreds of millions of times across an aging world. Death is no fun at the best of times. But death in such Dickensian circumstances and in such numbers is a numbing thought.

Months later, I visited the occupants of a Red Cross home for destitute old men in a poor and dangerous corner of Panama City. When we think of the old, we usually think of women, because there are more of them. But for many men who reach old age, the outlook is even worse than for their womenfolk. Especially when they have no family.

There were about sixty of them, each over sixty years old, in a single dormitory. Each had a bed and a small metal locker for his possessions. Most of them watched the TV in the corner, the only other piece of furniture. "We feed them, we clothe them, and we bury them," sighed Ada Guzmet, my female guide. She said this without malice, but with a good deal of cynicism. Most of these men had families, she said. But "they got thrown out of their homes. They were male chauvinist pigs when they were younger. Macho. Now their wives don't want them and their children don't want them. Only their mothers wanted them, and their mothers are gone, so they have nobody left. Nobody comes to visit them."

Actually, much of this seemed untrue. There were often other reasons why they were estranged from their families. Antonio Idalgo was eighty-eight years old and had been in the home for twenty of them. He was proud of his English, learned eight decades ago at the city's Pan-American school, where his father had taught. Later Antonio worked as a bookkeeper for Chiriqui Land, a banana company. Living on the company plantations, "I became a stranger to my family," he told me. But he was no passive victim of circumstances. "I have jobs. I sell chewing gum and sweets on the street and spend my money on girlfriends and lottery tickets."

Close by, Juan-Paulo was perched on a high stool with his shirt hanging out, reading comics through a magnifying glass. "He's always reading," said Ada. Juan-Paulo had been in the home for only two years. "I used to work up-country for a vineyard," he said. "I mapped the boundaries in the mountains and lived on the vineyard. I am here because they knocked down my home when I stopped working there."

There was a waiting list to enter the home, but Ada said it was hard finding the money to keep it open. They held rummage sales to try to raise money. "But people won't give money for the elderly like they do for children, and especially not for old men." The one solace the men had was their girlfriends. Most claimed to have one or more on the go. Maybe so. There are plenty to go around. Women live to eighty in Panama, six years longer than men. But Ada wondered how genuine these women were. "Women from the Dominican Republic come here. They try and trick the men into marrying them so they can get papers and stay in Panama. Things are better here than there."

Already just over half the world's old people live in developing countries. By 2050 there will be 1.5 billion people aged over sixty-five round the world, and 1.2 billion of them will be in what is now the poor world. "Older people are the poorest in every developing country," says Sarah Harper of the Oxford Institute of Ageing. "They consistently have the lowest levels of income, education, and literacy. They lack savings, assets, and land, have little skill or capital to invest in productive activity, and have only limited access to labor, pensions, or other benefits." Many are also sick and alone. With few, if any, children to look after them in old age, they are victims, you might say, of their acquiescence in government policies to reduce family sizes.

Demographers say wistfully that there may be only one way out: euthanasia. In China today, doctors openly discuss it. Its newspapers report that *anlesi,* or "tranquil death," is widespread in hospitals. Is this just a matter of allowing people in pain to die, or is it something more? In 2007 a headline in the *China Daily* asked, "Merciful or Ruthless? Plea for euthanasia chills China." Margaret Sleeboom-Faulkner of the University of Leiden has warned that "the weakness of health care provision for the terminally ill in China" is bound to lead to "violations." Will China's leaders one day feel compelled to make euthanasia compulsory, just as their predecessors once found compulsory birth control necessary?

Equally worrying is the suggestion of futurologist Jesse Ausubel that our older societies, stripped of their youthful vigor, may simply give up and shrivel away, like some African tribes confronted with the modern world. Homo sapiens could go out not with a bang but with an incontinent whimper.

Or is there an alternative? Is there a silver lining? Yes.

CHAPTER 27

Silver Lining

In 1965 the Who sang that they hoped to die before they got old. Today, those rock stars who survived the drug binges and the fast cars are old, and often still rocking–making more use of condoms than colostomy bags. Mick Jagger remains a sex symbol at an age that would once have rendered him infirm. He is nobody's idea of an aging "dependent," even if that is how the demographic data categorize him. Old women can have it too. In early 2009, Tina Turner took to the stage in London, dancing in heels, microskirts, and a gown slashed to the navel. In her seventieth year. And it's not just the rock stars. As Guy Brown, a researcher on longevity at Cambridge University, noted with what sounded like salivating desire, "Dame Helen Mirren, at sixty-three, shares little or nothing with a centenarian in the dementia unit of a care home."

The old are ever more active, assertive, and independent. And their numbers seem likely to reorder society to make this ever more possible. They fill libraries and seminar halls once crammed with callow youths. When I give lectures, some of my best audiences are not students but feisty grays with the skills to sustain an argument that they must have learned on the campuses of the sixties. They run picket lines and marathons, take up skateboarding, and play the stock market. Arguably, the old are not a threat to future societies at all, but an opportunity—a new resource to be tapped.

Demographers talk a lot about how the oldies are creating soaring "dependency ratios." About how there will be ever more old people to be cared for with the cash earned by ever fewer working adults. But this is a one-sided analysis. Demographers are stuck in statistical tram lines of their own making. They forget that while there may be more and more old people to support, there are fewer young children to look after. They forget that we now benefit from a hugely increased economic role for women in society. They forget that we lose far less time to illnesses these days. All these factors increase society's ability to look after the old. But above all, demographers are prone to forget the changing nature of the old themselves. They need far less looking after and can contribute more to society.

Some smart people saw this coming. When Quaker activist Maggie Kuhn set up the Gray Panthers in the United States in 1972, she spotted the radical potential of society's seniors: "The old, having the benefit of life experience, the time to get things done, and the least to lose by sticking their necks out, are in a perfect position to serve as advocates for the larger public good."

In China, most of the frequent demonstrations by aggrieved workers in rust belt industrial towns are organized by old people demanding back pensions from failing privatized companies. The state holds few fears for these truculent old heroes from another era. They have seen it all, from the '49 revolution through the Great Leap Forward of the 1950s to the cultural revolution of the '60s and the new capitalist revolution. Many of them were among Mao's Red Guards and remember when workers were revered and their factories were shining new socialist enterprises. As one younger protester put it in admiration, "Older workers are not afraid. They see no difference between starving to death and being killed."

The longevity revolution has a long way to go. Medical researchers expect that a life expectancy of ninety or more will soon become the norm as the health of the elderly continues to improve. Those at an age regarded today as dependent, doddering, and feebleminded will be in future be wise, wicked, and working. Or if not working, then doing the host of family, community, and social activities once performed by housewives.

This extension of the working and contributing lives of society's most qualified, most experienced individuals is potentially a huge new demographic resource. Like the youth bulge before it, the silver bulge can be seen as either a threat or a promise. Fear it, and our societies will founder; harness it, and the prospects are endless. Millions of middle-class retirees already continue working at everything from lucrative consultancies to serving in the local charity shop—frequently both. Often they are much more valuable to society than the young workers that demographers imagine are supporting them.

The idea of a retirement age was invented by Otto von Bismarck in the 1880s when he needed a starting age for paying German war pensions. It was done cynically. He chose sixty-five because his officials told him it was the typical age at which old soldiers died. But it seems increasingly a policymakers' whim. Men can currently expect to have a retirement of more than twenty years, and women can expect one approaching thirty years. Many want to work longer today, but are stopped.

Many governments, including the British, are reassessing the idea of enforced retirement. One day soon, old people will be required to work longer. And why not? British journalist Katherine Whitehorn, herself eighty but still pounding away at the laptop, puts it this way: "The days when you were at work by your early 20s at latest, rose to some sort of seniority by 45, and retired at 60 or so are obviously over. It seems perfectly obvious that the old are going to have to work longer, and I would be astonished if anyone under 60 thought anyone older should be able to just sit back and put their feet up at their expense."

Some grumpy old men and women may fret at being forced to continue work. But they should volunteer, in return for laws to ban the ageism that blights the working lives of many in late middle age. They should campaign for a new deal for the old, in which society no longer marginalizes them but instead redesigns homes and cities, workplaces and work hours, to meet their physical and aesthetic requirements.

Some worry that an older workforce will be less cutting-edge, less adaptable. "Significant innovations and big discoveries tend to be made between the ages of 30 and 50, with peak creativity between 35 and 40," says demographer Nicholas Eberstadt. As societies age, "the critical groups of those of innovation age will decline." But research often shows that companies with a decent proportion of older workers are more productive than those addicted to youth. Age brings experience and wisdom that complements the attributes of the young. "Think what it will mean," says one academic, "when the Edisons and Einsteins of the future, the doctors and technicians, the artists and engineers, have twenty or thirty more years of life to give us."

Most elderly in Europe and North America resemble Ushi, the

Japanese centenarian described at the beginning of the last chapter, more than they resemble the wasting woman I met in Bangladesh or the men living in the Red Cross home in Panama City. Only one in twenty lives in a care home. Typically, the elderly are assets and not liabilities. They are the council members and counselors, the social secretaries and neighborhood wardens, the caregivers for other elderly people and even the agitators—the glue that holds busy societies together.

That is increasingly the case in some developing countries too. I was struck by the sudden importance that old people can have in poor societies when I visited South African townships for HelpAge International. The HIV virus has ripped through the urban heartlands of South Africa. With a quarter of adults infected, life expectancy has fallen from sixty to under fifty years. The funerals of young parents are a daily occurrence. They have left behind more than a million orphans. The conventional population pyramid is being turned into an hourglass, with plenty of the old and young but few people in between. So it is grandparents who look after the children. Tens of thousands of grandparents have nursed their own dying children and are now left holding the babies. One community nurse, Veronica Khosa, told me, "Of the one hundred and eighty HIV-positive families I am working with right now, over a hundred are run by grandparents."

Lembe was living in Ennerdale, a township outside Johannesburg. At sixty-eight, she could barely walk, even with a stick. She had cared for all three of her daughters as they fell sick from AIDS. Now they were gone. She was still paying off the debt for the funeral of the last daughter to die. Meanwhile, she was looking after three granddaughters and no fewer than six grandsons in a tiny three-room shack. The grandchildren lined up on the sofa to heap praise on her. Any misgivings, I asked? Yes, said one of the grandsons. "Granny can't play football."

Maria, in her mid-seventies, was living in Mamelodi, north of Pretoria, with her daughter Margaret. They were looking after four of Maria's grandchildren and four great-grandchildren, three of whom were left behind by Margaret's two daughters. Both the daughters had died of AIDS in the two years before my visit. The fathers of the grandchildren lived locally but provided no support. One was HIV positive and probably dying. Nobody had a job. This menagerie of ten people from four generations all lived in Maria's house. Their main income was Maria's old-age pension, plus rent from letting out a shack at the back of her house.

Finally, I met Molly and her husband in Elim, close to the border with Zimbabwe. Aged seventy-two and sixty-three, they lived with two daughters, one clearly mentally ill, plus the daughters' three children and five more grandchildren, the offspring of two other daughters who had died of AIDS. Again, their only income was the state pension. "You can't throw the children out. It's what God gives you," said Molly. Lembe, Maria, and Molly dependent? Give me a break. Their society couldn't do without them.

Old people are consumers as well as caregivers. Back in the rich world, the "silver market" is huge. Anybody watching U.S. network TV will know the constant advertising aimed at the elderly, from Viagra to retirement homes. Oldies have savings in banks and pension funds, and cash from the sale of large houses they no longer need. The money is all available for purchases and investment. While Japan's domestic economy was close to a standstill through the 1990s, the country remained the world's largest exporter of capital—thanks in considerable part to pensioners' savings, earned during the boom years of Japan's own manufacturing industry. Those savings were now funding steelworks in Malaysia, forestry projects in Brunei, mines in Indonesia, and electronics factories in China. It is not too fanciful to suggest that it could be the savings of the old that rescue the post-recession global economy. And the wealth of the oldies will all ultimately be available for the next generation—either to family members or, through state inheritance taxes, to society in general.

In the end, the rise of the old is not about economics or retirement ages or the elderly as caregivers. It is about society's zeitgeist, its cultural and social wellsprings. Theodore Roszak, a historian at California State University and author of *The Longevity Revolution*, is one of the most passionate and effective advocates for the virtues of aging. He once wrote me a splendidly acid letter after I suggested in a magazine article that an aging society might be a society in trouble. "We ought to begin recognizing longevity as the greatest collective benefit yet to emerge from the Industrial Revolution," he wrote. "Aging is the best thing that has happened in the modern world, a cultural and ethical shift that looks a lot like sanity." On reflection, I think he is right. Many of our basic nostrums will change with the rise of the wrinklies, Roszak says. "This is terra incognita. Our species has never lived in societies where there were more people above the age of 50 than below." With tribal elders in the majority.

This story is full of ironies. The baby boomers of the second half of the twentieth century, the people who brought the cult of youth to new heights, seem about to pioneer the age of the old. The generation that brought us Woodstock and the hippies, the Beatles and Bill Clinton, may indeed have been special, spanning the crucial years of the people-quake and determining its progress. Roszak, arch exponent now of the virtues of the coming age of the old, was himself one of the baby boomers' luminaries and chroniclers, with past books like *The Making of a Counter Culture*, published in 1968. Now sporting his own gray hairs, he declares that the boomers "brought the global industrial hegemony to its climax. . . . Now it is their destiny to herald the new senior dominance."

In nomadic societies, the old were simply left behind to die. In agrarian societies, they were kept on to tend the land. In industrial societies, they were bundled into old folks' homes. In our aging post-industrial world, the elders will for the first time be in charge. The implications will be profound. Roszak sees a more caring world. "The old are not a good audience for a dog-eat-dog social ethic," he says. "If anything, they create an ambience which favors the survival of the gentlest." That must be good news.

The older we are, the less we are hooked on the latest gizmos—whether military toys or PlayStations—and on obsessive consumption. The older we are, the more we seem to appreciate the finer things in life and the things that last. We may, in consequence, reduce pressure on the world's resources by consuming less and by paying greater attention to ensuring our air is cleaner, our biodiversity richer, our soils more fertile, and our climate more predictable.

The Commission on Global Aging worries that in an older society, "armed forces may experience chronic manpower shortages" and "elder-dominated electorates may be more risk-averse, shunning decisive confrontations abroad." But that rather presumes that the era of the old should want to behave as if it were still in its testosterone-driven twenties. A few fewer hormones in international affairs, a few more "decisive confrontations" shunned, might be good thing. More sensibly, the commission remarks that "at 50 [individuals] do not expect to act or behave or feel as we did at 20—nor at 80 as we did at 50. The same is true for entire societies. What will it be like to live in societies that are much older than any we have ever known or imagined?" If it is anything like the Japanese centenarians' community of Ogimi, it looks fit, frugal, and just fine.

CHAPTER 28

Peak Population and Beyond

Even before most people had heard about the population explosion, the number of children being born to the average women in the world was falling. Fertility peaked at between five and six children per woman in the 1950s. During that decade, the fraction of the world's population under five years old reached 15 percent. Since then, having babies has been going out of fashion, and growing old has been the new thing. Today an average woman has 2.6 children, and the fraction of the world's population under five is below 10 percent.

Smaller families did not immediately cut population growth. Rising life expectancy saw to that—along with the baby boomers' reaching adulthood and starting families themselves. The percentage rate at which the world's population was growing did not peak until the late 1960s. Joel Cohen, most eminent of demographers, says that "our descendants will look back on the 1960s peak as the most significant demographic event in the history of the human population, even though those of us who lived through it did not recognise it at the time." The peak rate of population growth was 2.1 percent. Since then, it has fallen to below 1.2 percent.

Percentages are not the same as absolute numbers. So some will give more historical significance to 1987, when the number of additional people on the planet each year peaked. That year ended with 87 million more people on the planet than it started with. Today, each year still adds 78 million, the equivalent of an extra America every four years. But that figure, too, is now on a near-inevitable downward slide, probably to zero and on into negative territory by midcentury. The final shock of the peoplequake is likely to be falling world population.

There have already been days when the world got smaller. On December 26, 2004, a quarter of a million died in the Indian Ocean tsunami. Added to the normal daily death toll of about 160,000, that made more than 400,000 deaths—easily exceeding the 370,000 daily new births. Some individual countries are now shrinking from year to year. In 2008 there were twenty-six of them, headed by Russia, Montenegro, Bulgaria, Zimbabwe, Ukraine, Latvia, and Swaziland. Many more will follow.

It took around 130 years, from about 1800 to 1927, for the world to get from one billion people to two billion, but only another thirty-three years to reach three billion, which happened in 1960. Reaching four billion took just fifteen years, until 1975. The fifth billion came in twelve years, in 1987, as did the next billion, achieved in 1999. The momentum created by a large young adult population of childbearing age is still pushing up population. Even so, the next billion people on the planet will take a little longer than the last—probably fourteen years, if we reach seven billion as expected in 2013. And getting to eight billion could take another twenty or more years, if we get there at all. Wolfgang Lutz of the Vienna Institute of Demography sees a peak as early as 2040, at closer to seven than eight billion, following by a strong downward slide, taking us to as low as five billion by 2100.

Peak population is probably much closer than most people think. A declining global population by later this century looks increasingly inevitable. Falling fertility means that we will soon reach the point where each succeeding generation of mothers will be smaller than the last. The demographic momentum will then be negative rather than positive. If each generation has only 1.6 babies to replace two adults, then five women are producing only four women for the next generation. Girls that are never born cannot have babies. The number of fertile females falls and falls and falls.

The stage is set for population peak and decline. Europe is into negative momentum already. Its native population could be halved by midcentury. By 2100, on current fertility trends, Germany could have fewer natives than today's Berlin, and Italy's population could crash from 58 million to just 8 million. Even if fertility recovers to about 1.85 children per woman, Ukraine will lose 43 percent of its population, Bulgaria and Georgia 34 percent, Belarus and Latvia 28 percent, and Romania, Russia, and Moldova each more than 20 percent.

Once the trend has set in, it will be very hard to break. As well as having ever fewer potential mothers, societies may get out of the habit of having babies. Children will be rare, exotic, and unusual. We can see this already. Only a few years ago, going to a café in Italy would see you surrounded by noisy children. Now you will likely see only adults, including many young latte-sipping men and women who would once have been surrounded by kids.

The world is destined to deal with the repercussions of the baby boom and subsequent bust as they play out over the coming decades. One of the most controversial consequences of the peoplequake is the already rising tide of migration. It is created in part by the obscene income differentials round the world, but even more by current record fertility differences. When some countries have more than six children per woman and others barely more than one, importing and exporting people is an obvious safety valve for both sides. Europe and East Asia already badly need foreign hands to keep their societies and economies functioning. They should stop pretending otherwise.

That movement of people—along with longer-established migration into North America—sounds like a good thing to me. Migrants are a major means of redistributing the world's wealth from the rich to the poor. Initially, most migrants do low-paid work like flipping hamburgers or turning sheets, looking after the young, the sick, and the elderly, harvesting fruit and vegetables, or cleaning trains and toilets. But increasingly they will do skilled jobs, becoming doctors and engineers and lawyers and civil servants. Migration is often seen as a symptom of disturbance and injustice round the world. It often is. But just as the migrants who fled Europe in past centuries created a "New World" in America, so migration in the twenty-first century may help remake the planet in beneficial ways. If America can succeed as a country of migrants, why can't every country?

The other critical change is aging. The youth bulge will be replaced by a silver bulge. But unlike the youth bulge, it will not be a passing phenomenon. It will be permanent, even after the baby boomers of the twentieth century die off. As fertility falls, the world is going to become older and older.

When the peoplequake subsides and the demographic plates settle down later this century, the world will be very different. We could have trashed our planet, unless we take serious steps to conserve natural resources and maintain a stable climate. At the worst, we could pass tipping points in climate change beyond which there is no recovery. But the optimist in me says that we will recognize the crisis for what it is and seize the moment to rein in our consumerism and reengineer industrial societies to avoid the worst environmental hazards.

We are coming through the greatest surge of human population numbers in our history. The peoplequake has already changed us profoundly, and its end game will change us even more. The reproductive revolution unleashed huge forces of economic activity, social dislocation, and liberation—for women in particular. Well before the end of this century, *Homo sapiens*—the brash, go-getting, hormone-driven young naked ape of the twentieth century—will be older and will likely be more conservative, less innovative, more boring even. But that is no bad thing. We need a breather. A stable, sagacious society that has lost its adolescent restlessness and settled into middle age sounds appealing.

Our world will be more densely populated for sure, but it will also be less frenetic and hopefully more humane—a kinder, gentler, wiser, and greener world. It probably won't be quite like the utopia conjured up by Malthus's old adversary, William Godwin: "A people of men and not children with no war, no crimes, no government . . . every man seeking the good of all." But except perhaps for his wish to see an "eclipsing of the desire for sex," we can hope.

Our species has already seen three great population surges in its history—each of them accompanied by technological innovation that increased the number of people the planet could support. First came tool making, then agriculture, and most recently industrialization. They have taken us from ten million people ten thousand years ago to almost seven billion now. But in between, there have been long periods of relatively stable world population. Now it looks like we are moving back into a more stable period. But we are permanently transformed. It is not just our technology that has changed. Our demography has changed too.

In the past, these stable eras all involved high death rates, high birth rates, and male dominance. Now we have the chance of a low-mortality, low-fertility future. And an end to patriarchy. The tribal elders may take center stage once more. But this time they will not just be revered, they will be the largest group in society. And in all probability, they will be dominated by women. Demographically, and probably in almost every other respect, it will be a very different world. As I grow older, it seems to me that it might be a much better world. But for good or ill, it is going to unfold over the coming century. There is no going back.

NOTES ON SOURCES

This is a partial list of personal, written, and electronic sources, concentrating on the main ones I used in the book, but including some others that I think will be particularly useful for students or are available most simply via the Internet.

I have interviewed dozens of people specifically for this book and thank them all. Most of them are acknowledged in the text or in the notes to individual chapters. But a number of demographers and other experts have had a wider influence on many chapters and my overall thinking. They include Joseph Chamie, Tim Dyson, Matthew Connelly, Joel Cohen, Betsy Hartmann, Gordon Conway, the late Julian Simon, Theodore Roszak, Lester Brown, David Satterthwaite, Michael Mortimore, Camilla Toulmin, Jim Lovelock, Jesse Ausubel, Misha Glenny, Vaclav Smil, Nicholas Eberstadt, Jack Caldwell, Peter McDonald, Zhongwei Zhao, John MacInnes, Lant Pritchett, Paul Ehrlich, Gavin Jones, and Wolfgang Lutz. I thank them in particular.

I should also thank the editors of *New Scientist, Yale Environment 360,* and the *Guardian,* as well as HelpAge International, WWF, Ohio State University, the International Institute for Applied Systems Analysis, New York University, the British Council in India, and the Australian Davos Connection for (knowingly or not) funding trips or commissioning articles that helped in the research for this book. None, of course, have any editorial responsibility for the content.

Where published sources are available through open Internet access, I have generally given a link. In addition, some references to news articles are currently available on free archive Web sites, including the *Guardian* and *Observer* (www.guardian.co.uk), the *New York Times* (www.nytimes.com/pages/world), and BBC News (http://news.bbc .co.uk).

Any readers wishing to find further references not covered below are welcome to contact me at pearcefred@hotmail.com.

CHAPTER 1: A DARK AND TERRIBLE GENIUS

The only proper biography of Malthus is *Population Malthus* by Patricia James (London: Routledge Kegan Paul, 1979). I have used it extensively, along with the different versions of Malthus's *Essay on the Principle of Population*, published from 1798. I also referred to an 1830 essay by Malthus, "A Summary View of the Principle of Population," found in *Three Essays on Population* (New York: Mentor, 1960). Among online sources I consulted are a review by John Avery on the Web site of the Danish Peace Academy (www.fredsakademiet.dk/library/avery/malthus.htm) and a handy short biography by Nigel Malthus at http://homepages.caverock.net.nz/~kh/ bobperson.html.

British parliamentary debates on the Poor Law reforms can be seen in *Hansard Parliamentary Debates*, for instance at http://hansard.millbanksystems.com/ lords/1838/mar/20/new-poor-law. For Marx, read *Marx and Engels on the Population Bomb*, edited by Ronald Meek (Berkeley: Ramparts Press, 1971). I also used some Marxist analysis from Eric Ross's *The Malthus Factor* (London: Zed Books, 1998), and an edited extract of this book published online by The Corner House at www .thecornerhouse.org.uk/pdf/briefing/20malth.pdf.

Information on the history of contraception came from *Montaillou* by Emmanuel Le Roy Ladurie (London: Scolar Press, 1978); Robert Engelman's *More: Population, Nature and What Women Want* (Washington, DC: Island Press, 2008); a paper by Etienne van de Walle and Virginie De Luca, "Birth Prevention in the American and French Fertility Transitions," *Population and Development Review* 32, no. 3 (September 2006): 529–55; "History of Contraception" by Malcolm Potts and Martha Campbell, in *Gynecology and Obstetrics* CD-ROM, edited by J.J. Sciarra (Philadelphia: Lippincott Williams & Wilkins, 2003); and Zhongwei Zhao's "Towards a Better Understanding of Past Fertility Regimes: The Ideas and Practice of Controlling Family Size in Chinese History," *Continuity and Change* 21, no. 1 (May 2006): 9–35.

John Reader wrote about foundling hospitals in *Cities* (London: Vintage, 2004). Francis Place's main work was *Illustrations and Proofs of the Principle of Population*, published most recently by Allen & Unwin (London, 1930). His life is also discussed by James in *Population Malthus* (cited above); in E. P. Thompson's *The Making of the English Working Class* (London: Victor Gollancz, 1980), which also has things to say about Malthus; and by John Caldwell in "The Delayed Western Fertility Decline: An Examination of English-speaking Countries," *Population and Development*

Review 25, no. 3 (September 1999): 479–513, which also discusses Comstock. Other valuable texts I consulted on this era include Robert Woods's *The Demography of Victorian England and Wales* (Cambridge: Cambridge University Press, 2000) and *Victorian Cities* by Asa Briggs (London: Penguin, 1968).

CHAPTER 2: THE ROAD TO SKIBBEREEN

My main source here was *The Great Hunger: Ireland 1845–1849* by Cecil Woodham-Smith (London: Hamish Hamilton, 1962), still the best study of the crime. Also valuable have been Christine Kinealy's book *The Great Calamity* (Dublin: Gill and Macmillan, 2006) and the work of Cormac O'Grada, in his book *Black '47 and Beyond* (Princeton, NJ: Princeton University Press, 2000) and in his articles available online, such as "Ireland's Great Famine," *ReFRESH* 15 (Autumn 1992), http://irserver.ucd.ie/dspace/bitstream/10197/418/3/ogradac_article_pub_079.pdf. I also consulted Eric Ross's *The Malthus Factor* (London: Zed Books, 1998) and John Newsinger's essay "The Great Irish Famine: A Crime of Free Market Economics," *Monthly Review*, April 1996, http://findarticles.com/p/articles/mi_m1132/is_n11_v47/ai_18205165/?tag=content;col1.

I used reports from newspaper archives, including the *Cork Examiner* archives at http://adminstaff.vassar.edu/sttaylor/FAMINE/Examiner/Archives, plus Skibbereen's own Web site (www.skibbereen.ie/famine.php). Other contemporary accounts consulted include *On the Causes of Distress at Skull and Skibbereen* by W. Neilson Hancock (Dublin: Hodges and Smith, 1850), www.tara.tcd.ie/bitstream/2262/7761/1/jssisiVolII1_10.pdf; James Mahoney's narrative in "The Irish Potato Famine, 1847," posted on the Web site EyeWitness to History at www.eyewitnesstohistory.com/irishfamine.htm; and John Henry Parker's *Narrative of a Journey from Oxford to Skibbereen* (Oxford, UK: John Henry Parker, 1847), http://adminstaff.vassar.edu/sttaylor/FAMINE/Journey/Frontispiece.html. The emaciated corpses are described by a local teacher and quoted in Cathal Poirteir's *Famine Echoes* (Dublin: Gill and Macmillan, 1995).

I also referred again to Patricia James's *Population Malthus* (London: Routledge Kegan Paul, 1979) and to Mike Davis's *Late Victorian Holocausts* (London: Verso, 2001), which lays bare how Malthusian thinking pervaded the British Empire.

CHAPTER 3: SAVING THE WHITE MAN

The early development of eugenics is covered in *A Life of Sir Francis Galton* by Nicholas Wright Gillham (Oxford, UK: Oxford University Press, 2001). His relationship with his cousin Charles Darwin is discussed in *Darwin* by Adrian Desmond and James Moore (London: Penguin, 1992). The Galton Institute provided useful material, including a brief biography of Julian Huxley by John Timson at www.galtoninstitute.org.uk/Newsletters/GINL9912/julian_huxley.htm.

Sidney Webb is quoted in *The Task of Social Hygiene* by Havelock Ellis (London: Constable, 1912). Get a flavor of Karl Pearson's writing at Paul Halsall's Internet Modern History Sourcebook, www.fordham.edu/halsall/mod/1900pearsonl .html. The thoughts of Churchill and others are in G. Searle's *Eugenics and Politics in Britain 1900–1914* (Leyden, Netherlands: Noordhoff, 1976), and those of others in Searle's "Eugenics and Politics in Britain in the 1930s," *Annals of Science* 36 (1979): 159–69. Blacker's role and other snapshots appear in Matthew Connelly's *Fatal Misconception* (Cambridge, MA: Belknap, 2008). Verschuer is discussed by Gregory Gardner and Maurice King in "A Martian View of the Hardinian Taboo," *British Medical Journal* 316 (1998): 1386.

John and Pat Caldwell assessed the influence of eugenics on demographers in *Limiting Population Growth* (London: Frances Pinter, 1986). A wealth of material about Margaret Sanger can be found at the Margaret Sanger Papers Project, www .nyu.edu/projects/sanger/aboutmspp/index.html. Some of the darker side emerges in feminist and antiracist texts such as Tanya L. Green's "The Negro Project: Margaret Sanger's Eugenic Plan for Black Americans," posted May 1, 2001, on the Web site of Concerned Women for America, www.cwfa.org/articledisplay.asp?id=1466& department=CWA&categoryid=life. Read the sad story of Ota Benga in *Ota Benga: The Pygmy in the Zoo* by Phillips Verner Bradford and Harvey Blume (New York: St. Martin's Press, 1992).

CHAPTER 4: AN ORNITHOLOGIST SPEAKS

European demography in the early twentieth century is discussed by Massimo Livi-Bacci in "Demographic Shocks: The View from History," a paper presented at the Federal Reserve Bank of Boston's forty-sixth economic conference in June 2001 (www.bos.frb.org/economic/conf/conf46/conf46c1.pdf). See also John Caldwell's "The Western Fertility Decline," *Journal of Population Research* 23 (2006): 225, and "Paths to Lower Fertility," *British Medical Journal*, October 9, 1999.

The evolution of population theory in midcentury is described by John and Pat Caldwell in *Limiting Population Growth* (London: Frances Pinter, 1986). Sir George Handley Knibbs gave his predictions in *The Shadow of the World's Future* (reprint, New York: Arno Press, 1976). Theodore Schulz's Nobel lecture in 1979, "The Economics of Being Poor," appears at http://ca.geocities.com/econ_0909meet/schultz-lecture.html.

Our Plundered Planet by Fairfield Osborn was published in 1948 (London: Faber & Faber). But the majority of this chapter deals directly with the content of William Vogt's book *Road to Survival*, published in 1948 in the United States (New York: W. Sloane Associates).

CHAPTER 5: THE CONTRACEPTIVE CAVALRY

Ishimoto Shizue and her times are profiled by Elise Tipton in "Ishimoto Shizue: The Margaret Sanger of Japan," *Women's History Review* 6 (1997): 337, and by J. Axelbank in "Japan's Family Planning Pioneer," *Populi* 15 (1988): 55.

Much of the documentation behind this chapter comes from a detailed archive search by the historian Matthew Connelly for his controversial book *Fatal Misconception* (Cambridge, MA: Belknap, 2008). His conclusions may not always be the same as mine, but his research is impeccable. John and Pat Caldwell corroborate much of what Connelly says in their book *Limiting Population Growth* (London: Frances Pinter, 1986) and elsewhere. See also Paul Ehrlich's "Demography and Policy: A View from Outside the Discipline," *Population and Development Review* 34, no. 1 (March 2008): 103–13.

Other useful sources include the Population Council Web site (www.pop council.org) and historical material provided by Pathfinder International (formerly the Pathfinder Fund), set up by Clarence Gamble (www.pathfind.org/site/PageServer?pagename=About_History). Also see Amartya Sen's essay "Population: Delusion and Reality," *New York Review of Books*, September 22, 1994.

Ansley Coale and Edgar Hoover published their arguments about driving the demographic transition in *Population Growth and Economic Development in Low-Income Countries* (Oxford, UK: Oxford University Press, 1959). Robert McNamara's alarm call linking aid to population control appeared in his book *One Hundred Countries, Two Billion People* (London: Praeger, 1973).

CHAPTER 6: THREE WISE MEN

Paul Ehrlich's original article on the population bomb, "Paying the Piper," appeared in *New Scientist*, December 14, 1967. His book, *The Population Bomb,* was published in New York by Ballantine Books in 1969. Among other sources on his work, I drew on "Betting on the Planet" by John Tierney, *New York Times*, December 2, 1990; Ehrlich's later books, notably *The Population Explosion*, written with his wife, Anne Ehrlich (London: Arrow, 1990); and a couple of interviews with him over the years.

Caldwell made his skeptical comments on Ehrlich's rhetoric and the limits-to-growth model in an article called "Man and His Futures" in the journal of the Australian and New Zealand Association for the Advancement of Science, *Search* 16 (February–March 1985): 22. Harriet Presser remembered the idea of needing a license to have children during a seminar I attended in Barcelona in July 2008, "Fertility and Public Policies in Low-fertility Countries," organized by the International Union for the Scientific Study of Population (www.iussp.org/Activities/low2/reportbarcelona08.pdf).

Garrett Hardin's essay "Lifeboat Ethics" appeared in *Psychology Today* in September 1974. Much of his writings can be found at the Web site of the Garret Hardin Society, www.garretthardinsociety.org, and at the Web site "Stalking the Wild Taboo," www.lrainc.com/swtaboo/index.html. "Tragedy of the Commons" appeared in *Science* 162 (1968): 1243.

The Limits to Growth by Donella Meadows, Dennis Meadows, et al., appeared in 1972 under the imprint of Universe Books, New York. Its launch was reported in

"The Limits to Growth: Hard Sell for a Computer View of Doomsday" by Robert Gillette, *Science* 175, issue 4026 (March 10, 1972): 1088–92. The analysis by the *New Internationalist* appeared as "The Shape of Things to Come?" (March 1972).

CHAPTER 7: SIX DOLLARS A SNIP

Matthew Connelly's *Fatal Misconception* is an important source of the Indian narrative. He summarized his thoughts in "Population Control in India: Prologue to the Emergency Period," *Population and Development Review* 32, no. 4 (December 2006): 629–67. Another mine of demographic information is *Twenty-First Century India*, edited by Tim Dyson et al. (New Delhi: Oxford University Press, 2004). A report on "Kerala's Pioneering Experiment in Mass Vasectomy Camps" by S. Krishnakumar appeared in the Population Council's journal, *Studies in Family Planning* 3, no. 8 (August 1972): 177.

Caldwell's recollections on the Indian emergency appeared in "A Polemic against Control," *Science* 321 (2008): 1043. Victorian famine stories come from Mike Davis's *Late Victorian Holocausts* (London: Verso, 2001). India's food situation in midcentury has been outlined by Lester Brown in *Seeds of Change* (New York: Praeger, 1970).

CHAPTER 8: GREEN REVOLUTION

The two paragraphs that open the chapter are reportage from my visit to India in 2004. A good primer on the green revolution and what needs to come now is Gordon Conway's *The Doubly Green Revolution* (London: Penguin, 1997). John McNeill wrote about the political context of the green revolution in his magisterial environmental history of the twentieth century, *Something New under the Sun* (London: Allen Lane, 2000). Eric Ross's *The Malthus Factor* (London: Zed Books, 1998) provides a spirited political attack on the revolution. More history is available at the Web sites of the various green revolution research centers, including the International Rice Research Institute at www.irri.org, CIMMYT (the International Maize and Wheat Improvement Center) at www.cimmyt.org, and the Consultative Group on International Agricultural Research at www.cgiar.org.

Julian Simon's important books include *The Ultimate Resource* (Princeton, NJ: Princeton University Press, 1981) and *The Resourceful Earth* with Herman Kahn (Oxford, UK: Basil Blackwell, 1984). My article on Simon's ideas appeared in *New Scientist* under the title "In Defence of Population Growth" on August 9, 1984, three days after his op-ed "The Myth of Over-Population" appeared in the *Wall Street Journal*.

Simon's bet with Paul Ehrlich is chronicled in an article called "Betting on the Planet" by John Tierney in the *New York Times*, December 2, 1990. Ester Boserup's *The Conditions of Agricultural Growth* was published in 1965 by Allen & Unwin,

London. She also published a memoir considering the work's significance three decades later, titled *My Professional Life and Publications 1929–1998* (Copenhagen: Tusculanum Press, 1999).

CHAPTER 9: ONE CHILD

The story of Mao and Ma is told in *Mao's War Against Nature* by Judith Shapiro (Cambridge: Cambridge University Press, 2001). James Lee and Wang Feng analyze the non-Malthusian demography of pre-Mao China in their paper "Malthusian Models and Chinese Realities," *Population and Development Review* 25, no. 1 (March 1999): 33–65. Vaclav Smil analyzed "China's Great Famine: 40 Years Later" in the *British Medical Journal*, December 18, 1999.

China's coercive population policies have been followed with vigor by journalists such as Jasper Becker (for instance, "Chairman Mao's Bleak Bequest," *Guardian*, February 23, 1990) and Liu Yin ("China's Wanted Children," *Independent*, September 11, 1991). More recently, they have been examined by Sing Pao in *Oriental Daily*, May 21, 2007 (www.zonaeuropa.com/20070521_2.htm) and by Tania Branigan in "Days of the One-Child Rule Could Be Numbered," *Guardian*, February 29, 2008, and "Chinese Babies Seized for Sale to Overseas Families," *Guardian*, July 4, 2009.

Matthew Connelly's *Fatal Misconception* uncovered the International Planned Parenthood Federation's duplicitous position on China. I quoted Emil Salas and Jack Caldwell from Mexico City in "UN Defends Tactics for Population Control," *New Scientist*, August 9, 1984. Nathan Keyfitz's "The Population of China" appeared in *Scientific American* in February 1984. Linda Chalker's 1995 parliamentary answer on Qian Xinzhong can be read at the Hansard Web site, http://hansard .millbanksystems.com/written_answers/1995/oct/30/unfpa-population-award-to-dr-qian.

Most of the comments from Zhongwei Zhao, plus recent demographic data, come from an interview with him in 2008. I have also consulted his papers "Re-examining China's Fertility Puzzle," *Population and Development Review* 32, no. 2 (June 2006): 293–321, and "Long-Term Mortality Patterns in Chinese History," *Population Studies* 51 (1997): 117, as well as his book with Fei Guo, *Transition and Challenge: China's Population at the Beginning of the 21st Century* (Oxford, UK: Oxford University Press, 2007). Also see Xizhe Peng's 2004 commentary, "Is It Time to Change China's Population Policy?" in *China: An International Journal* 2, no. 1 (March 2004): 135–49, http://muse.jhu.edu/journals/china/v002/2.1peng.pdf.

Betsy Hartmann wrote about Bangladesh's about-face on contraception in *A Quiet Violence* with James Boyce (London: Zed Press, 1983) and *Reproductive Rights and Wrongs* (Boston: South End Press, 1995).

CHAPTER 10: SMALL TOWNS IN GERMANY

The material on Hoyerswerda and the surrounding area was gathered during a visit to the town in 2008, guided by anthropologist Felix Ringel. I also conducted interviews with Mathias Siefhoff, a geographer at the Technical University in Dresden, and colleagues Peter Wirth and Joachim Ragnitz. I consulted the city's Web site (www.hoyerswerda.de/city_info/webaccessibility/index.cfm?waid=34), as well as Stefan Skora's maps at www.lanu.de/media/files/Akademie/PDF/Gewaessertage/ skora_stefan.pdf and statistics at the Lower Saxony government Web site, www .statistik.sachsen.de.

Reiner Klingholz of the Berlin Institute for Population and Development (www .berlin-institut.org/about-us.html) provided data, including the institute's reports *The Demographic State of the Nation: How Sustainable Are Germany's Regions?* (2006) and *Male Emergency* (2007). I consulted the proceedings of the International Symposium on Coping with City Shrinkage and Demographic Change, held in Dresden in 2006 (see www.schader-stiftung.de). Jean-Claude Chesnais's paper "Below-Replacement Fertility in the European Union" appears in *Review of Population and Social Policy* no. 7 (1998): 83–101.

Some descriptions of social conditions in eastern Germany come from press reports, notably in the *New York Times* ("A Wave of Attacks on Foreigners Stirs Shock in Germany," October 1, 1991; "As East Germany Rusts, Young Workers Leave," December 25, 2002; "Last Out, Please Turn Out the Lights," May 28, 2004; "In Eastern Germany, an Exodus of Young Women," November 8, 2007); in the *Guardian* ("Exodus from East Leaves Land of Broken Promises," November 15, 2006; "Slow Death of a Small Germany Town," January 27, 2008); in *Deutsche Welle* ("No Brakes on Germany's Population Freefall," August 17, 2006); and in *Der Spiegel* ("Racist Manhunt in Small-Town Germany," August 21, 2007; "Wolves Solidify Paw-Hold in Germany," October 26, 2007).

CHAPTER 11: WINTER IN EUROPE

Early demographic fears in Europe were reported in "Europe's Baby Bust," *Newsweek*, December 15, 1986, and in a resolution passed in the European Parliament (Resolution C127/78) on April 12, 1984.

Major sources for this chapter included papers and discussions at the International Seminar on Fertility and Public Policies in Low Fertility Countries, held in Barcelona on 7–9 July 2008 by the International Union for the Scientific Study of Population (IUSSP) (www.iussp.org/Activities/low2/reportbarcelona08.pdf). Other major sources were "Lowest-low Fertility in Europe" by Hans-Peter Kohler et al., published in *The Baby Bust: Who Will Do the Work? Who Will Pay the Taxes?* edited by F. R. Harris (Lanham, MD: Rowman & Littlefield, 2006) and available online at www.ssc.upenn.edu/~hpkohler/papers/Low-fertility-in-Europe-final.pdf; and "Demographic Change and Response: Social Context and the Practice of Birth Control in Six Countries" by Harriet Presser et al., *Journal of Population Research* 23 (2006): 135.

I also drew on correspondence with Massimo Livi Bacci, author of *A Concise History of World Population* (Chichester, UK: Wiley, 2006), and Maria-Letizia Tanturri of the University of Florence, as well as her paper "Childless or Childfree?"

in *Population and Development Review* 34, no. 1 (March 2008): 51–77. Wolfgang Lutz wrote "The Low-Fertility Trap Hypothesis," *Vienna Yearbook of Population Research* 2006, p. 167, and "Europe's Population at a Turning Point," *Science* 299, no. 5615 (March 28, 2003): 1991–92. I read *When Mothers Work and Fathers Care* by Gosta Esping-Andersen et al. (DemoSoc Working Paper 2005–5, Universitat Pompeu Fabra, Barcelona, May 2005, www.recercat.net/bitstream/2072/2036/1/ DEMOSOC5.pdf) and Peter McDonald's "Low Fertility and the State," *Population and Development Review* 32, no. 3 (September 2006): 485–510, http://paa2005 .princeton.edu/download.aspx?submissionId=50133.

There are illuminating debates on the Demography Matters blog (http://demog raphymatters.blogspot.com), such as the April 2008 post by Edward Hugh on "Famil iarism in Italy" (http://demographymatters.blogspot.com/search?q=familiarism).

Other sources on Italy include Gianpiero Dalla Zuanna et al., "Low Fertility and Limited Diffusion of Modern Contraception in Italy," *Journal of Population Research* 22 (2005): 21, and Alessandra De Rose et al., "Italy: Delayed Adaptation of Social Institutions to Change in Family Behaviour," *Demographic Research* 19 (2008): 665–704.

Here, as elsewhere, I have generally taken current fertility rates from estimates available on the CIA World Fact Book Web site (https://www.cia.gov/library/ publications/the-world-factbook).

Wassilios Fthenakis made his comments about the demographic impact of the financial crisis to the German public broadcaster ARD; they were reported by Reu ters on February 17, 2009. Carl Haub considered the likely impact in a blog post called "Will the Economic Downturn Lower Birth Rates?" on the Web site of the Population Reference Bureau on January 8, 2009 (http://prbblog.org/?p=34). David Reher wrote "Towards Long Term Population Decline," *European Journal of Population* 23 (2007): 189, and discussed his work at the Demography Matters blog on April 19, 2008 (http://demographymatters.blogspot.com/2008/04/global-demographics-almost-uncharted.html).

CHAPTER 12: RUSSIAN ROULETTE

Boris Yeltsin's antics are all over the media; see, for instance, video on YouTube at www.youtube.com/watch?v=Q5FIoocja4k. Nicholas Eberstadt surveys the wreck age in *Russia's Demographic Disaster*, a special report published by the American Enterprise Institute (Russian Outlook series, May 2009, www.aei.org/outlook/ 100037). See also Tim Helcniak's 2002 article, "Russia's Demographic Decline Con tinues," on the Population Reference Bureau's Web site, www.prb.org/Articles/ 2002/RussiasDemographicDeclineContinues.aspx. Michael Beliak discussed "The Baby Deficit in Russia," *Science* 312 (2006): 1894.

Anatoly Vishnevsky wrote on "The Specter of Immigration" in *Russia in Global Affairs*, no. 2, April-June 2005 (http://eng.globalaffairs.ru/numbers/11/915.html). Paul Demeny's Yemen comparison appears in "Population Policy Dilemmas in

Europe," *Population and Development Review* 29, no. 1 (March 2003): 1–28. I read Vladimir Putin on "Raising Russia's Birth Rate" in *Population and Development Review* 32, no. 2 (June 2006): 385–89. Michael Ryan wrote on alcoholism, the price of sausages, and Russia's rising mortality in the *British Medical Journal*, March 11, 1995.

Journalists whose work I used here are Luke Harding, "No Country for Old Men," *Guardian*, February 11, 2008, and Graeme Smith, "Russia Shrinks," *Globe and Mail* (Toronto), April 22, 2006. "Romania Presses for Procreation" appeared in *New Scientist* on December 15, 1986. Maire Ni Bhrolchain and Tim Dyson wrote about Romanian demography in "On Causation in Demography," *Population and Development Review* 33, no. 1 (March 2007): 1–36. I interviewed Marek Nowak of the Polish Sociological Association in Poznan in late 2008, and read his and other contributions to *Declining Cities, Developing Cities*, edited by Nowak and Michal Nowosielski (Poznan, Poland: Instityut Zachodni, 2008).

CHAPTER 13: SISTERS

I met Akhi, Aisha, and Miriam, as well as Mashuda Shefali Khatun, in Dhaka in 2007. John Caldwell investigated "The Bangladesh Fertility Decline" in *Population and Development Review* 25, no. 1 (March 1999): 67–84. Mizanur Rahman discussed the fertility decline in *When Will Bangladesh Reach Replacement-Level Fertility? The Role of Education and Family Planning Services*, an internal paper for the UN Population Division in 2003 (http://huwu.org/esa/population/publications/com pletingfertility/2RevisedRAHMANpaper.PDF). This paper was part of a wider internal exercise by the UN Population Division, "Completing the Fertility Transition," available at www.un.org/esa/population/publications/completingfertility/completingfertility.htm.

Spirited support for family planning programs comes from "A Response to Critics of Family Planning Programs" by John Bongaarts and Steven Sinding, in *International Perspectives on Sexual and Reproductive Health* 35, no. 1 (March 2009). But see also *Too Many People?* by Nicholas Eberstadt (report published by the International Policy Network, London, July 2007, www.aei.org/docLib/20070712_Too_Many_People.pdf).

I reported on the Cairo population conference in 1994 for *New Scientist* (for instance, "Women's Rights Dominate Cairo Plan," September 17, 1994). And Phyllida Brown contributed "Sisters Are Doing It for Themselves," *New Scientist*, August 20, 1994.

I discussed Iran with Amir Mehryar at the IUSSP seminar in Barcelona (see notes for chapter eleven). And I interviewed Soraya Tremayne in London in early 2009. Her paper "Modernity and Early Marriage in Iran: A View from Within" appeared in *Journal of Middle East Women's Studies* 2, no. 1 (Winter 2006): 65–94, http://muse.jhu.edu/journals/journal_of_middle_east_womens_studies/v002/2.1tremayne.pdf. Read about the country's condom factory in "Condoms Help Check Iran Birth

Rate" by Jim Muir, April 24, 2002, on the BBC News Web site at http://news.bbc
.co.uk/1/hi/world/middle_east/1949068.stm.

See also Mohammad Jalal Abbasi-Shavazi's paper "Recent Changes and the
Future of Fertility in Iran" on the UN Web site (http://huwu.org/esa/population/
publications/completingfertility/2RevisedABBASIpaper.PDF) and his paper with
Amir Mehryar, Gavin Jones, and Peter McDonald, "Revolution, War and Modernization: Population Policy and Fertility Change in Iran," *Journal of Population
Research* 19, no. 1 (March 2002): 25–46. Peter McDonald wrote on Iran in *Fertility and Contraceptive Use Dynamics in Iran: Special Focus on Low Fertility Regions*
(Working Paper No. 1, Australian Demographic and Social Research Institute,
2007, http://adsri.anu.edu.au/pubs/ADSRIpapers/ADSRIwp-01.pdf).

Contraception overviews derive from *Recent Levels and Trends of Contraceptive
Use as Assessed in 1983* (New York: United Nations, 1984), and regular updates appear
in the United Nations Population Division series *World Contraceptive Use.*

Gavin Jones took an important look at "A Demographic Perspective on the Muslim World," *Journal of Population Research* 23 (2006): 243. A major International
Union for the Scientific Study of Population conference on low fertility in East and
Southeast Asia took place in Tokyo in November 2008 (www.iussp.org/Activities/
low2-index.php).

CHAPTER 14: SEX AND THE CITY

If you have missed *Sex and the City*, check out the show's Web site, www.hbo.com/
city. In the more likely event you have missed *The Balzac Age*, try www.siberian
light.net/the-balzac-age-or-all-men-are-basta. Chen Chang told her story in "New
Women, Old Problems in China" by Holly Williams, November 13, 2002, on the
BBC News Web site at http://news.bbc.co.uk/1/hi/world/asia-pacific/2460087
.stm.

More young women in cities? Read Lena Edlund's "Sex and the City," *Scandinavian Journal of Economics* 107 (2005): 25, http://papers.ssrn.com/sol3/papers
.cfm?abstract_id=703310. The man-drought narrative is in Bernard Salt's book *Man
Drought* (Melbourne: Hardie Grant Books, 2008).

John MacInnes's texts on the reproductive revolution continue to evolve. I have
read and discussed with him several versions, including "The Reproductive Revolution," a paper written with Julio Pérez and presented at the International Union
for the Scientific Study of Population's twenty-fifth International Population Conference in Tours, France, in July 2005 (www.ced.uab.es/publicacions/PapersPDF/
Text270.pdf).

Peter McDonald wrote "Gender Equity, Social Institutions and the Future of
Fertility," *Journal of Population Research* 17, no. 1 (May 2000): 1–16, and "Low Fertility and the State: The Efficacy of Policy," *Population and Development Review*
32, no. 3 (September 2006): 485–510, http://iussp2005.princeton.edu/download

.aspx?submissionId=50830. His compatriot John Caldwell looks at "Demographic Theory: A Long View" in *Population and Development Review* 30, no. 2 (June 2004): 297–316. Reportage on the same theme comes from the BBC, "Gender Issues Key to Low Birth Rate," November 20, 2007, http://news.bbc.co.uk/1/hi/world/asia-pacific/7096092.stm.

Tim Dyson spoke to me at length about Indian women. And I read S. Padmandas on "Compression of Women's Reproductive Spans in Andhra Pradesh, India" in *International Family Planning Perspectives* 30, no. 1 (March 2004): 12, www.guttmacher.org/pubs/journals/3001204.html.

Tom Espenshade's 2006 work at Princeton University's Office of Population Research on replacement fertility is found in his paper "How Many Children Does It Take to Replace Their Parents? Variation in Replacement Fertility as an Indicator of Child Survival and Gender Status," presented at the annual meeting of the Population Association of America in Los Angeles, March 2006, and available at http://paa2006.princeton.edu/download.aspx?submissionId=60125.

CHAPTER 15: SINGAPORE SLING

My main interviewees in Singapore were Daniel Goh, Gavin Jones, and Mui Teng Yap. I made extensive use of articles published in the *Strait Times* during my visit. For instance: "New Citizens Make Up Half Our Olympic Team," "Love 101 Proves Top Hit in Class," and "Bigger Role for Dads," all on August 18, 2008; "More Help with Bringing Up Baby," "PM Lee Recalls His Nappy-Changing Days," "Let Children Learn at Their Own Pace," and "Why Not Paternity Leave, Ask Couples," all on August 19, 2008; and "Babies: Call It a Crisis" and "The Men Who Don't Get It," both on August 20, 2008. "No Money, No Honey" appeared in the *Straits Times* on April 20, 2009.

I received polling and other material from the Working Mothers Forum in Singapore. Liberal intellectuals who "harbor homosexuality" were "unmasked" in 2006 in "Co-Ed Schools Versus Single Sex Schools," posted at a ThinkQuest Web site dedicated to education policy in Singapore, http://library.thinkquest.org/05aug/01348/coed.html. And I checked out dating Web sites like www.romancingsingapore.com and www.lovebyte.org.sg.

Academic material consulted included Mui Teng Yap's 2007 paper, "Fertility and Population Policy: The Singapore Experience," *Journal of Population and Social Security (Population)*, supplement to vol. 1, p. 643, and Theresa Wong and Brenda Yeoh's *Fertility and the Family: An Overview of Pro-natalist Population Policies in Singapore* (Asian MetaCentre Research Paper Series No. 12, National University of Singapore, 2003, www.populationasia.org/Publications/RP/AMCRP12.pdf). Lutz explored Singapore's education system in "Reconstructing the Past," in the newsletter of the International Institute for Applied Systems Analysis, *Options*, in summer 2006.

I discussed the politics of baby bonuses with Peter McDonald and others. Carl

Haub is quoted in "No Babies," *New York Times*, June 29, 2008. Tim Dyson's conclusion that men need to become more like women first appeared in an article I wrote on low fertility, "Mamma Mia," in *New Scientist*, July 20, 2002.

CHAPTER 16: MISSING GIRLS

There was a flurry of concern in the Mumbai media about "missing girls" during my visit in August 2008; for example, "Six More Diagnostic Clinics in City Raided," *Mumbai Mirror*, August 28, 2008; "Continuing Mystery of Mumbai's Missing Girls" and "Illegal Sonography Machines Sealed at Malpani's Clinic," both *Times of India*, August 28, 2008. The BBC reported the rare case of a doctor arrested for performing a sex-selective abortion in "India Sex Selection Doctor Jailed," March 29, 2006 (http://news.bbc.co.uk/1/hi/world/south_asia/4855682.stm).

Population First's website is at www.populationfirst.org. Studies of the India "missing girls" that I consulted include "Sex Ratio at Birth and Excess Female Child Mortality in India" by Perianayagam Arokiasamy, presented at a seminar on "Female Deficit in Asia: Trends and Perspectives," held in Singapore in December 2005 by the Center for Population and Development. R. L. Bhat and Namita Sharma's paper "Missing Girls: Evidence from Some North India States" appeared in *Indian Journal of Gender Studies* 13 (2006): 351. The Haryana court judgment was reported in March 2006 on the Web site of the Indian Ministry of Health and Family Welfare, http://mohfw.nic.in/dofw%20website/acts%20&%20rules/Judgment_Palwal.htm. A book on the issue, *Watering the Neighbour's Garden: The Growing Demographic Female Deficit in Asia*, edited by Isabelle Attané and Christophe Z. Guilmoto (Paris: Committee for International Cooperation in National Research in Demography, 2007), is available online at www.cicred.org/Eng/Publications/pdf/BOOK_singapore.pdf.

I discussed Korea's problems with Minja Kim Choe at the IUSSP seminar in Barcelona (see notes for chapter eleven), and read her work, such as her 2007 paper "How Does Son Preference Affect Populations in Asia?" (*Asia-Pacific Issues* no. 84, September 2007, www.eastwestcenter.org/fileadmin/stored/pdfs/api084.pdf). The issue is also discussed by UN demographer Joseph Chamie in a blog, "The Global Abortion Bind," posted on the Web site of the Yale Center for the Study of Globalization on May 29, 2008 (http://yaleglobal.yale.edu/display.article?id=10886), and by Sylvie Dubuc and David Coleman in "An Increased Sex Ratio of Births to Indian-Born Mothers in England and Wales," *Population and Development Review* 33, no. 2 (June 2007): 383–400.

The Chinese missing-girls problem has been discussed by Judith Banister in "Shortage of Girls in China Today," *Journal of Population Research* 21, no. 1 (May 2004): 19–45, http://findarticles.com/p/articles/mi_moPCG/is_1_21/ai_n6155263. William Lavely presented his paper "Explaining the Female Deficit in a Chinese County" in Singapore in December 2005, at the International Conference on the Female Deficit in Asia, organized by the National University of Singapore

(www.cicred.org/Eng/Seminars/Details/Seminars/FDA/PAPERS/39_Lavely
.pdf). Other papers at that conference included "A Sharp Increase in Sex Ratio at
Birth in the Caucasus" by France Mesle et al.

CHAPTER 17: WHERE MEN STILL RULE

During my visit to Israel, I had long interviews with Menachem Friedman, Akiva
Wolff, Carmi Wisemon, Sergio DellaPergola, and Danny Seidmann; and, among
the Bedouin, Amal Elsana-Alh'jooj and Alean Al-Krenawi. They provided most of
the information here.

The Web site of the Israeli newspaper *Haaretz*, at www.haaretz.com, was also
valuable on the Haredi community. It included the stunning report on Ahuva
Klachkin, "How Many Children Does It Take to Be Righteous?" by Tamar Rotem,
November 20, 2006 (www.haaretz.com/hasen/spages/789928.html). Inescapable,
too, was the Web site of the Ettinger Report by lobbyist Yoram Ettinger at http://
yoramettinger.newsnet.co.il/Front/NewsNet/newspaper.asp; likewise the anony-
mous paper "The Fertility Dynamic of Israel's Ultra-Orthodox Jews and Pronatalist Government Policy" that I found at the Web site of Focus Anthropology,
a peer-reviewed publication of undergraduate research (www.focusanthro.org/
archive/2005-2006/katz0506.pdf).

I also consulted Marwan Khawaja's paper "The Fertility of Palestinian Women
in Gaza, the West Bank, Jordan and Lebanon," *Population-E* 58 (2003): 273; Philippe
Fargues's "Protracted National Conflict and Fertility Change: Palestinians and
Israelis in the Twentieth Century," *Population and Development Review* 26, no. 3
(September 2000): 441–82; the Web site of the Negev Institute for Strategies of
Peace and Development (www.nisped.org.il); and press reports on the 2008 Pales-
tinian census, including "Census Finds Palestinian Population Up by 30 Per Cent,"
Guardian, February 11, 2008; "Haredi Growth," *Jerusalem Post*, November 9, 2005;
and "The Demographic Bomb Is a Dud," *Jerusalem Post*, January 14, 2005. "Man-
aging Migration: The Global Challenge" by Philip Martin and Gottfried Zürcher,
Population Bulletin 63, no. 1 (March 2008), was also useful.

CHAPTER 18: WAVING OR DROWNING?

The grisly fate of the Chinese cockle pickers was widely reported in the British
media. Among the most useful articles were "Going Under," *Guardian*, June 20,
2007; "One Desperate Man's Path to His World's End," *Times* (of London), March
25, 2006; "Victims of the Sand and the Snakehead," *Guardian*, February 7, 2004;
and "Two Days Before the Disaster, Yu Phoned His Wife," *Guardian*, February 20,
2004.

Hsiao-Hung Pai wrote many articles for the *Guardian* about Chinese migrants,
such as "Another Morecambe Bay Is Waiting to Happen," March 28, 2006; "'Local
Men Have a Special Liking for Foreign Girls,'" April 22, 2008; and "'If You Care
about Your Family, Give Us £12,000,'" April 23, 2008. Much of this material was

included in her book *Chinese Whispers* (London: Penguin, 2008). Frank Pieke, director of Chinese studies at Oxford University, has studied the Fujian people trade in *Chinese Globalization and Migration to Europe* (Working Paper 94, Center for Comparative Immigration Studies, San Diego, March 2004).

The *Guardian* has also reported well on boat people. For instance, see "Washed Up on the Beach," May 10, 2006; "Over 70 Migrants Feared Killed on Crossing to Europe," August 28, 2008; and "Greek Islands Become the EU's New Front Line on Immigration," October 17, 2008. Also see "After 20 Days Adrift," *Guardian*, February 4, 2009.

Other press reports I consulted include "Thais 'Leave Boat People to Die,'" BBC News Web site, January 15, 2009 (http://news.bbc.co.uk/1/hi/world/south_asia/7830710.stm); "Massacre at Sea," *Observer*, January 19, 1997; "Libya Key Transit for UK-Bound Migrants," *Observer*, January 13, 2008; and "Lampedusa Swept by Migrant Landings," *Ansa*, June 23, 2008.

Here and in the next chapter I have made extensive use of reports by Hein de Haas, such as *Turning the Tide?* (Working Paper 2, International Migration Institute, University of Oxford, 2006) and "The Myth of Invasion," *Third World Quarterly* 29, no. 7 (2008): 1305–1322. Both are available on his Web site at www.heindehaas.com. On trans-Saharan migration, I consulted a study by Martin Baldwin-Edwards, *Migration in the Middle East and Mediterranean* (Geneva: Global Commission on International Migration, 2005, www.mmo.gr/pdf/news/Migration_in_the_Middle_East_and_Mediterranean.pdf).

Italian battles have been reported by the *Guardian*: "Bitter Harvest," December 19, 2006; "68% of Italians Want Roma Expelled," May 17, 2008; "Italy Sends Troops into Camorra's Heartlands after Mafia Killing of Migrants," September 24, 2008; "Apartheid in the Heart of Europe," November 16, 2007; and "Polish Labourers Kept in Italian Prison Camp," July 20, 2006. *L'espresso* also carried articles that I consulted in English, such as "I Was a Slave in Puglia," September 4, 2006.

The World Bank reported usefully on Europe's need for migrant labor in *Shaping the Future* in March 2009. Peter McDonald and Rebecca Kippen discussed "The Implications of Below Replacement Fertility for Labour Supply and International Migration 2000–2050" at a meeting of the Population Association of America in Los Angeles in March 2000 (http://dspace.anu.edu.au/bitstream/1885/41469/8/labourpaper.pdf).

Gregory Maniatis was writing for the openDemocracy Web site, in an article titled "The Road to Nowhere," October 5, 2005 (www.opendemocracy.net/people migrationeurope/migration_2898.jsp). A manifesto for a fairer European migration policy appears in Oxfam's report *Walls on the Sea* (2007).

CHAPTER 19: MIGRANT MYTHS

My London neighbors are identified from an interactive map on the *Guardian* Web site (www.guardian.co.uk/flash/0,,1398066,00.html). Also consulted to dispel the myths were "Migration to European Countries: A Structural Explanation of Patterns, 1980–2004" by Marc Hooghe et al., *International Migration Review* 42 (2008): 476; "Toward a Demography of Immigrant Communities and Their Transnational Potential" by Jorgen Carling, *International Migration Review* 42 (2009): 449–75; and proceedings of the Conference on Migration, Development, and Pro-poor Policy Choices in Asia, held in Dhaka, Bangladesh, in June 2003 (papers from www.livelihoods.org).

I read "Challenges and Opportunities—The Population of the Middle East and North Africa" by Farzaneh Roudi-Fahimi and Mary Mederios Kent, *Population Bulletin* 62, no. 2 (June 2007), and Wolfgang Lutz on *The Contribution of Migration to Europe's Demographic Future* (report of the International Institute for Applied Systems Analysis, September 2007, www.iiasa.ac.at/Admin/PUB/Documents/IR-07-024.pdf).

Stephen Castles is quoted from "Development and Migration—What Comes First?" presented in February 2008 in New York City at a conference called "Migration and Development: Future Directions for Research and Policy," organized by the Social Science Research Council (www.imi.ox.ac.uk/pdfs/S%20Castles%20Mig%20and%20Dev%20for%20SSRC%20April%202008.pdf). Other papers by Castles that I read included *Environmental Change for Forced Migration* (Working Paper No. 70, New Issues in Refugee Research series, UN High Commission for Refugees, October 2002, www.unhcr.org/research/RESEARCH/3de344fd9.pdf). His book with Mark Miller, *The Age of Migration*, 4th ed. (New York: Guildford Publications, 2009), is also an important reference.

Misha Glenny's remarkable book *McMafia* (London: Bodley Head, 2008) has many stories about people trafficking. The plight of Saudi house slaves is reported in *As If I Am Not Human*, a report published by Human Rights Watch in 2008. The story of Brazilian slaves is told in "Brazilian Taskforce Frees More Than 4,500 Slaves," *Guardian*, January 3, 2009.

Many remittance statistics come from the World Bank's Migration and Development Brief in 2008 (http://blogs.worldbank.org/peoplemove/outlook-for-remittance-flows) and from papers such as Maria Cristina Pantiru's *Migration and Poverty Reduction in Moldova* (Sussex Center for Migration Research, February 2007). Brenda Yeoh and Mohammed Mizanur Rahman wrote on *hundiwalas* in *The Social Organization of Remittances* (Asian MetaCentre Research Paper Series No. 20, National University of Singapore, 2006, www.populationasia.org/Publications/RP/AMCRP20.pdf).

Hein de Haas (see notes for last chapter) also wrote "International Migration Remittances," *Third World Quarterly* 26 (2005): 1269. Ronald Skeldon wrote "International Migration as a Tool of Development Policy," *Population and Development Review* 34, no. 1 (March 2008): 1–18. Also interesting was "Feminization of Migration," a 2006 article by Nancy Yinger on the Population Reference Bureau Web site (www.prb.org/Articles/2006/TheFeminizationofMigration.aspx).

I interviewed Lant Pritchett in Cambridge in October 2008. I read his paper written with Michael Clemens, "Income Per Natural: Measuring Development for People Rather Than Places," *Population and Development Review* 34, no. 3 (September 2008): 395–434, and another written with Clemens and Claudio Montenegro, *The Place Premium* (Working Paper No. 148, Center for Global Development, July 2008), plus an online interview, "Ending Global Apartheid" (February 2008) at the Web site of *Reason* magazine, www.reason.com.

Among Joseph Chamie's online articles are "Dying to Get In: Global Migration" (October 11, 2005) and "International Migration and the Global Agenda" (June 23, 2008), both on the Globalist Web site, www.theglobalist.com. Material on the U.S.-Mexican border came from Douglas Massey's "Backfire at the Border" in the Cato Institute's *Trade Policy Analysis* no. 29, June 13, 2005 (www.freetrade.org/pubs/pas/tpa-029.pdf); "The Cat and Mouse Game at the Mexico-U.S. Border" by Katharine Donato et al., *International Migration Review* 42, no. 2 (June 2008): 330–59; "In the Hands of People Smugglers," *Guardian*, July 2, 2008; and Charles Bowden's article "Exodus," *Orion*, July 2008.

CHAPTER 20: FOOTLOOSE IN ASIA

Gavin Jones was an important source for this chapter, including our interview and several papers, notably "Delayed Marriage and Very Low Fertility in Pacific Asia," *Population and Development Review* 33, no. 3 (September 2007): 453–478, and The *"Flight from Marriage" in South-East and East Asia* (Asian MetaCentre Research Paper Series No. 11, National University of Singapore, 1993, www.populationasia .org/Publications/RP/AMCRP11.pdf).

I also interviewed Mika Toyota, a researcher of Japanese migrants, at the National University of Singapore. She alerted me to the Lucky Plaza mall in Singapore, where I met Bridget. Brendah Yeoh wrote on Singapore maids in "Transnational Domestic Workers in Global-City Singapore," an article on the National University of Singapore's Research Gallery Web site at www.fas.nus.edu.sg/rg/html/geog/byshindex. html. Nicholas Eberstadt wrote about the plight of the grown-up little emperors in his book *Prosperous Paupers and Other Population Problems* (Edison, NJ: Transaction Publishers, 2000).

Naohiro Ogawa and others wrote "Late Marriage and Less Marriage in Japan," *Population and Development Review* 27, no. 1 (March 2001): 65–102. The breakdown of conventional marriage is explored in the *Guardian* ("Go Home and Multiply, Japanese Told," November 29, 2008) and by William Sparrow in Bangkok, "When Freaky-Deaky Equals Hara-Kiri," *Asia Times*, March 8, 2008, www.atimes .com/atimes/Japan/JC08Dh01.html. More soberly, there is "Social Networks and Family Change in Japan" by Ronald Rindfuss et al., *American Sociological Review* 69 (2004): 838.

Kent Ewing blogs on Hong Kong for *Asia Times* (www.atimes.com/atimes/China/IH09Ad02.html). The Humanitarian Organization for Migration Economics (HOME) is at http://home.org.sg. The travails of its shelter were reported in the *Straits Times*, August 19, 2008. The organization Transient Workers Count Too has a Web site at www.twc2.org.sg. I also interviewed its director, John Gee.

David McKenzie wrote "A Profile of the World's Young Developing Country International Migrants," *Population and Development Review* 34, no. 1 (March 2008): 115–35. Some material came from a conference on "The Female Deficit in Asia" held in Singapore in December 2005, including a paper by Wen Shan Yang and Ying-ying Tiffany Liu on "Gender Imbalances and the Twisted Marriage Market in Taiwan." Malaysia invites rich foreigners at the "Malaysia My Second Home" Web site, www.mm2h.gov.my. "The Bar Girl and the Expat: A Killing Foretold" by Ian MacKinnon and Andrew Drummond appeared in the *Observer*, August 17, 2008.

CHAPTER 21: GOD'S CRUCIBLE

I gathered quite a lot of material for this chapter from the displays at the Ellis Island Immigration Museum, which I recommend (www.ellisisland.org). I also took the tour at the Tenement Museum in October 2008 (www.tenement.org), as well as visiting neighborhoods like Chinatown, Williamsburg, Little Italy, and Brighton Beach. *How the Other Half Lives* by Jacob Riis (reprint, New York: Dover Books, 1971) is a much-treasured book of photojournalism.

The story about Leland Stanford comes from "Chinese Transformed Gold Mountain," *Sacramento Bee*, January 18, 1998. I also read "Immigration: Shaping and Reshaping America" by Philip Martin and Elizabeth Midgley, *Population Bulletin* 61, no. 4 (December 2006); "America's Never-Ending Debate" by Douglas Massey, *Population and Development Review* 32, no. 3 (September 2006): 573–84; material on the Web site of the Pew Hispanic Center (http://pewhispanic.org); and Joseph Chamie's "12 Million Shadows: America's Immigration Dilemma," *Yale-Global*, February 4, 2008, http://yaleglobal.yale.edu/content/12-million-shadows-america%E2%80%99s-immigration-dilemma.

I dipped into Geoffrey Moorehouse's *Imperial City* (London: Hodder & Stoughton, 1988). More recent information on migration to New York comes from a paper called "The Impact of Immigration on New York City" by Peter Loby of the New York City Department of City Planning, Population Division. Data on where migrants live today comes partly from the Ellis Island exhibits and partly from "Immigrant Profiles of U.S. Urban Areas" by Lawrence Brown et al., *The Professional Geographer* 59 (2007): 56.

Check the Carrying Capacity Network's campaigns at www.carryingcapacity.org and the Council of Conservative Citizens at http://cofcc.org. They are illuminating if you hold your nose. Sierrans for U.S. Population Stabilization remain active at www.susps.org. I interviewed Betsy Hartmann on this topic for *New Scientist* in "The Greening of Hate," February 22, 2003. See also her contributions to a 2007–08 dialogue on "Population and Climate Change" on the *Bulletin of the Atomic Scientists* Web site, www.thebulletin.org/roundtable/population-climate-change; and "Documenting Racism in the Green Campaign Against Immigration," posted on the Web site of the Committee on Women, Population and the Environment, July

22, 2006, www.cwpe.org/node/145. (The latter is based on a document prepared by the Political Ecology Group, originally published in *Political Environments* 6, Fall 1998.) And check out "Anti-Migrants Plan Coup at 100-Year-Old Green Group" in the *Guardian*, January 23, 2004, and "Climate of Injustice," *New Scientist*, May 7, 2005.

I interviewed Hasan Omar in 2008 while visiting Ohio State University in Columbus. At the university, I also talked to Tamar Mott and read her paper "Immigration Profiles of U.S. Urban Areas and Agents of Resettlement," *The Professional Geographer* 59 (2007): 56, and *African Refugee Resettlement in the United States* (El Paso, TX: LFB Scholarly Publishing, 2009). Other sources include "A Somali Influx Unsettles Latino Meatpackers," *New York Times*, October 16, 2008; "Somalis Flee to World's Biggest Refugee Camp," *Guardian*, August 29, 2008; and "Fear of Change in Columbus," a paper by Kristin Shelley of the Columbus Metropolitan Library.

CHAPTER 22:
THE TIGERS AND THE BULGE

David Bloom, David Canning, and Jaypee Sevilla wrote about the demographic window (or dividend, it's the same thing) in *The Demographic Dividend* (Santa Monica, CA: RAND Corporation, 2003).

Wolfgang Lutz discussed the issue of the demographic window with me in an interview at the International Institute for Applied Systems Analysis in Laxenburg, Austria. I also read his papers and articles, such as "India's Window of Opportunity," *Options*, Autumn 2005, and "Population Dynamics of Human Capital Accumulation," in Prskawetz et al., "Population Aging, Human Capital Accumulation, and Productivity Growth," a supplement to *Population and Development Review* 34 (2008). In addition, I read Lutz's book *The End of World Population Growth in the 21st Century* (London: Earthscan, 2004). His ideas on literate life expectancy are articulated in different language in "The Mental Wealth of Nations" by John Beddington et al., *Nature* 455 (2008): 1057.

I interviewed Sanjeev Sanyal in Singapore. His book *The Indian Renaissance* was published in New Delhi by Viking, 2008. Ran Bhagat was speaking with me at the International Institute for Population Sciences in Mumbai. Lori Ashford wrote the policy brief *Africa's Youthful Population: Risk or Opportunity?* for the Population Reference Bureau, June 2007, www.prb.org/pdf07/AfricaYouth.pdf.

The Security Demographic, Population Action International's 2005 report, is available at www.populationaction.org/Publications/Reports/The_Security_Demographic/Summary.shtml. Richard Cincotta discussed the demographics of insurgency in the Worldwatch Institute's 2005 *State of the World* report (https://www.worldwatch.org/node/76). The intellectual origins of fears about a "youth bulge" may lie in a 1960s paper called "Youth as a Force in the Modern World" by Herbert Moller, *Comparative Studies in Society and History* 10 (1968): 237.

Joseph Chamie is quoted from our interview in New York in 2008. Jeffrey Sachs is quoted from an op-ed article, "No Place for Piety," in *New Scientist*, November

8, 2003. Thomas Homer-Dixon is quoted by Robert Kaplan in his 1994 *Atlantic Monthly* article, "The Coming Anarchy," later expanded into a book of the same name (New York: Vintage, 2002).

The U.S. National Security Strategy is at the Pentagon Web site, www.au.af .mil/au/awc/awcgate/nss/nss-95.pdf. I came across Gunnar Heinsohn's writings in the *Jerusalem Post*: "Youth Bulge Violence," April 10, 2007, and "Ending the West's Proxy War against Israel," January 29, 2009.

CHAPTER 23: FOOTPRINTS ON A FINITE PLANET

The Optimum Population Trust's Web site is at www.optimumpopulation.org. Lovelock made his "billion people" claim first in "James Lovelock: You Ask the Questions," *Independent*, August 14, 2006. Other recent discussions include Daniel Engber's "Global Swarming," *Slate*, September 10, 2007, www.slate.com/id/2173458.

A handy inventory of the state of the planet and an analysis of Ehrlich's equation appear in *AAAS Atlas of Population and Environment* by Paul Harrison and Fred Pearce (Berkeley: University of California Press, 2000), available online at www .ourplanet.com. Also see Paul Harrison's excellent *The Third Revolution* (London: IB Taurus, 1992). And for a guide to practical solutions, there are few better sources than Lester Brown's *Plan B 3.0: Mobilizing to Save Civilization* (New York: Norton, 2008).

Ecological footprint analysis is provided by the Global Footprint Network at www.footprintnetwork.org and in the annual *Living Planet* reports of WWF (www.panda.org/about_our_earth/all_publications/living_planet_report).

Stephen Pacala, director of the Princeton Environment Institute, reported on how the rich and poor worlds compare in their carbon dioxide emissions in "Equitable Solutions to Greenhouse Warming: On the Distribution of Wealth, Emissions and Responsibility Within and Between Nations," a lecture given in Vienna in November 2007 at a conference called "Global Development: Science and Policies for the Future," organized by the International Institute for Applied Systems Analysis (IIASA) (www.iiasa.ac.at/Admin/INF/conf35/docs/speakers/pacala.html). Costa Rica's story was also told at that conference by former environment minister Carlos Manuel Rodriguez (www.iiasa.ac.at/Admin/INF/conf35/docs/speakers/ rodriguez.html). I wrote up the conference in a report posted on the IIASA's Web site at www.iiasa.ac.at/Admin/PUB/policy-briefs/pb02-web.pdf.

China's real carbon emissions were revealed by Christopher Weber in "A Third of China's Carbon Footprint Blamed on Exports," *New Scientist*, July 30, 2008, www.newscientist.com/article/dn14412-33-of-chinas-carbon-footprint-blamed-on-exports.html. And China's footprint is compared with that of the United States in the Worldwatch Institute's 2006 *State of the World* report (www.worldwatch .org/node/3866). Chris Goodall's *How to Live a Low-Carbon Life* was published in London by Earthscan in 2007. A handy list of the carbon intensity of different national economies is presented on Wikipedia (http://en.wikipedia.org/wiki/ List_of_countries_by_ratio_of_GDP_to_carbon_dioxide_emissions).

CHAPTER 24: FEEDING THE WORLD

The story of the final land enclosures appears in John McNeill's *Something New under the Sun* (London: Allen Lane, 2000). Robert Zeigler told reporters about International Rice Research Institute research in April 2008, including Agence France Press (" 'Biofuels Frenzy' Fuels Global Food Crisis: Experts," April 29, 2008, http://afp.google.com/article/ALeqM5jnrykdFNonv92dF_OkNy5K1wHFQA). Josette Sheeran told the Associated Press about urban hunger the same month, as reported in "Rise in Food Prices Sparks Unrest" by Julien Spencer, posted April 2, 2008, on the *Christian Science Monitor* Web site, www.csmonitor.com/2008/0402/p99so1-duts.html. Chris Haskins was writing on "The Return of Malthus" in *Prospect*, January 2008, www.prospect-magazine.co.uk/article_details.php?id=9943. The commodities expert is quoted in "Malthusian Catastrophe Coming Soon, Says Commodities Expert Garry White" on the Fleet Street Invest Web site, June 5, 2008 (www.fleetstreetinvest.co.uk/commodities/fundamentals/malthusian-catastrophe-00050.html).

This chapter is informed by an excellent discussion of future food prospects, and also recent land grabs by rich countries in the developing world, in Alex Evans's report for Chatham House, *The Feeding of the Nine Billion* (London: Royal Institute of International Affairs, 2009, http://globaldashboard.org/wp-content/uploads/2009/Chatham_House_Feeding_Nine_Billion.pdf), and by discussions at a Wilton Park conference I attended in November 2008 called "Feeding the World" (www.wiltonpark.org.uk/documents/conferences/WP927/pdfs/WP927.pdf). Also important has been the report of the International Assessment of Agricultural Knowledge, Science and Technology for Development, chaired by the estimable Bob Watson, published in 2008 (www.agassessment.org). I also quote Watson from a meeting on agricultural renaissance held at Oxford University's James Martin 21st Century School in July 2009.

More on land grabs can be found in "Fears for the World's Poor Countries as the Rich Grab Land to Grow Food" by John Vidal, *Guardian*, July 4, 2009. For water shortages in agriculture, see my book *When the Rivers Run Dry* (Boston: Beacon Press, 2006). For a scary take on soils, I recommend *Dirt: The Erosion of Civilizations* by David R. Montgomery (Berkeley: University of California Press, 2007). Lester Brown wrote on "Starving People to Feed the Cars" in the *Washington Post*, September 10, 2006 (www.washingtonpost.com/wp-dyn/content/article/2006/09/08/AR2006090801596.html). Joel Cohen wrote on feeding ten billion, and much else, in an article called "Human Population Grows Up" in *Scientific American*, September 2005.

Jesse Ausubel described the impact of the world's adopting U.S. gain yields by 2050 in an interview with me; it is essentially repeated at the Rockefeller University Web site, http://phe.rockefeller.edu/SAF_Forest.

James Gasana's report on Rwanda for the International Union for the Conservation of Nature is repeated at the People and Planet Web site ("Remember Rwanda?" posted January 6, 2003, www.peopleandplanet.net/doc.php?section=2&id=1780).

Jared Diamond's book *Collapse* (New York: Viking, 2005) discusses the issue. Gerard Prunier's quotes come from his book *The Rwanda Crisis 1959–1994* (Kampala, Uganda: Fountain Publishers, 1999). The point is reiterated by Robin Mearns and Melissa Leach in *Lie of the Land* (Oxford, UK: James Currey, 1996). Robert Ford was writing on "Marginal Coping in Extreme Land Pressures" in *Population Growth and Agricultural Change in Africa*, edited by B. L. Turner II et al. (Gainesville: University Press of Florida, 1993).

Thomas Homer-Dixon wrote on Rwanda in a paper with Valerie Percival, *Environmental Scarcity and Violent Conflict: The Case of Rwanda* (Washington, DC: American Association for the Advancement of Science and the University of Toronto, 1995, www.library.utoronto.ca/pcs/eps/rwanda/rwanda1.htm). The conclusion is supported by BBC journalist Fergal Keane in his searing firsthand report from the conflict, *Season of Blood* (London: Viking, 1995).

Maurice King discusses Africa's crisis on his Web site, Demographic Entrapment, at www.leeds.ac.uk/demographic_entrapment, and in an unpublished paper, "The Demographic Entrapment of Middle Africa." John Guillebaud's quotes come from the 2007–08 dialogue on "Population and Climate Change" on the *Bulletin of the Atomic Scientists* Web site, www.thebulletin.org/roundtable/population-climate-change. Sachs wrote on Malawi's lessons for feeding Africa in "How to End the Global Food Shortage," *Time*, April 24, 2008, www.time.com/time/magazine/article/0,9171,1734834,00.html.

I have visited the Machakos district of Kenya twice in recent years and can testify to the truth of the analysis in *More People, Less Erosion* by Michael Mortimore, Mary Tiffen, and Francis Gichuki (Chichester, UK: Wiley, 1994). See also Mortimore's *Roots in the African Dust* (Cambridge, UK: Cambridge University Press, 1998). I visited the highlands of western Kenya in 2000.

Peter Holmgren reported on the afforestation of modern Kenya in *Ambio* 23 (1995): 390. Melissa Leach and James Fairhead wrote their analysis of West Africa's forests in "Demon Farmers and Other Myths," *New Scientist*, April 27, 1996, and subsequently developed the idea as "Challenging Neo-Malthusian Deforestation Analyses in West Africa's Dynamic Forest Landscapes," *Population and Development Review* 26, no. 1 (March 2000): 17–43.

Chris Reij outlined his ideas in an interview with me for *New Scientist* ("And Then There Were Trees," March 29, 2008), from which I quote. His thoughts also appeared in *Farmer Innovation in Africa: A Source of Inspiration for Agricultural Development* (London: Earthscan, 2001) and in "Unrecognised Success Stories in Africa's Drylands," *Haramata*, December 2007.

Norman Myers has written widely on environmental refugees (for instance, his paper "Environmental Refugees: An Emergent Security Issue," presented at the Organization for Security and Co-operation in Europe's Thirteenth Meeting of the Economic Forum in Prague, May 2005, www.osce.org/documents/eea/2005/05/14488_en.pdf). But see also Richard Black's *Environmental Refugees: Myth or Reality?* (Office of the United Nations High Commissioner for Refugees,

2001, www.unhcr.org/research/RESEARCH/3ae6a0d00.pdf); Fiona Flintan's "Environmental Refugees—A Misnomer or a Reality?" (paper presented at the Wilton Park Conference on Environmental Security and Conflict Prevention, Sussex, England, March 2001, www.ucc.ie/famine/GCD/Paper%20for%20Wilton%20Park.doc); and perhaps best of all, Cecilia Tacoli's "Crisis or Adaptation? Migration and Climate Change in a Context of High Mobility" (paper presented at a UN expert conference group meeting, "Population Dynamics and Climate Change," London, June 2009, www.unfpa.org/webdav/site/global/users/schensul/public/CCPD/Tacoli%20Abstract.pdf).

Catherine Maternowska called Haiti "an island of environmental refugees" in "Real Lives 1: Haiti," *People and the Planet* 3, no. 4 (1994): 16–19. Jared Diamond discusses the environmental history of Haiti in *Collapse* (New York: Viking, 2005). And see "The Evolution of the Haitian Diaspora in the USA" on the Haiti and the USA Web site (www.haiti-usa.org/modern/evolution.php).

CHAPTER 25: SLUMDOGS ARISE

I visited Mumbai, including Dharavi, for this book in the summer of 2008, where I interviewed, among others, Sheela Patel of the nongovernmental organization SPARC (the Society for the Promotion of Area Resource Centres). I am indebted for my thinking on cities over many years to David Satterthwaite of the London-based International Institute for Environment and Development (IIED), and for his publications such as *The Scale of Urban Change Worldwide 1950–2000* (London: IIED, 2005); *Squatter Citizen* (London: Earthscan, 1989) and *Empowering Squatter Citizen* (London: Earthscan, 2004), both with Diana Mitlin; and *Environmental Problems in an Urbanizing World*, with Mitlin and Jorge Hardoy (London: Earthscan, 2001).

Also valuable over the years have been Herbert Girardet's books, such as *Gaia Atlas of Cities* (London: Gaia Books, 1996), and John Reader's *Cities* (London: Vintage, 2004). Also, in their very different ways, I am indebted to Jane Jacobs's *The Death and Life of Great American Cities* (New York: Random House, 1961) and Oscar Lewis's *The Children of Sanchez* (London: Secker and Warburg, 1962), classics both. I have written about greening cities in "Ecopolis Now," *New Scientist*, June 16, 2006.

Future carbon emissions from urbanization were calculated by Michael Dalton et al. in "Demographic Change and Future Carbon Emissions in China and India," a draft of which was presented at the annual meeting of the Population Association of America in New York City in 2007 (www.iiasa.ac.at/Research/PCC/pubs/dememiss/Daltonetal_PAA2007.pdf). Joel Cohen's calculation of future urbanity is in an interesting interview with Jacques Gordon in *PREA Quarterly*, Summer 2006 (www.rockefeller.edu/labheads/cohenje/PDFs/328InterviewPREAQuarterly2006.pdf). His book *How Many People Can the Earth Support?* was published in New York by Norton, 1996.

My copy of Mike Davis's *Planet of the Slums* was published in London by Verso, 2006. Dharavi as "a vision of urban hell" comes from "Next Stop Squalor" by John Lancaster, *Smithsonian*, March 2007. Also see "Charles Declares Mumbai Shanty Town Model for the World," *Guardian*, February 6, 2009.

I have reported from many slums over the years, visiting Kibera (and Karen) in Nairobi on several occasions ("Back to Basics," *New Scientist*, May 27, 2000), as well as Orangi in Karachi in 1996 ("Squatters Take Control," *New Scientist*, June 1, 1996) and Rio de Janeiro's favelas in 1992 ("A Shanty Town That's Here to Stay," *New Scientist*, September 5, 1992).

CHAPTER 26: THE AGE OF THE OLD

Ushi Okushima's story is told by Kimiko de Freytas-Tamura in "Humanity's Oldest Village Fights to Stay Fit in a World of Unhealthy Habits," *Brunei Times*, March 23, 2007 (www.bt.com.bn/en/classification/life/features/2007/03/23/humanitys_ oldest_village_fights_to_stay_fit_in_a_world_of_unhealthy_habits), and by Jonathan Watts in "Japan's Island of Longevity," *Christian Science Monitor*, May 16, 2002 (www.csmonitor.com/2002/0516/p07s01-woap.html).

"Japanese Longevity" is discussed by Eileen Crimmins et al. in *Population and Development Review* 34, no. 3 (September 2008): 457–82. Also look at the site of the Okinawa Centenarian Society (www.okicent.org/team.html). Naohiro Ogawa wrote a paper titled *Rapid Population Aging in Japan* (Tokyo: Nihon University Population Research Institute, 2003), available at www.ancsdaap.org/cencon2003/ Papers/Japan/Japan.Ogawa.pdf. I read about Ogama village being put up for sale in "Aging and Official Abandonment Carries Japanese Village to Extinction," *New York Times*, April 26, 2006. See more on Japan's rural decline in "Where Have All the Young Men Gone?" *Economist*, August 24, 2006.

Julian Chapple wrote "The Dilemma Posed by Japan's Population Decline," published on the Web site of the *Electronic Journal of Contemporary Japanese Studies* on October 18, 2004, www.japanesestudies.org.uk/discussionpapers/Chapple .html. Vaclav Smil looked at "The Unprecedented Shift in Japan's Population" in the online magazine *Japan Focus*, posted on April 19, 2007 (www.japanfocus.org/ -Vaclav-Smil/2411).

I first wrote about global aging in *New Scientist* ("Welcome to the Global Old Folks' Home," July 9, 1987). The Commission on Global Aging published *Global Aging: The Challenge of the New Millennium* in 1999 (Washington, DC: Center for Strategic and International Studies, www.csis.org/files/media/csis/pubs/global aging.pdf), producing headlines like "Ageing Population Is Killing World Economy," *Guardian*, August 30, 2001.

Hunter-gatherers are discussed by Michael Gurven and Hillard Kaplan in "Longevity among Hunter-Gatherers," *Population and Development Review* 33, no. 2 (June 2007): 321–65. Ben Wattenberg's book *Fewer* was published in Chicago by Ivan R. Dee, 2004. Another primer is *The Age of Aging* by George Magnus (Chichester,

UK: Wiley, 2008). The United States' coming pensions crisis is discussed by Megan McArdle in "No Country for Young Men," *Atlantic Monthly*, January 2008.

I visited an area near Khulna in Bangladesh in 2007, and Panama City with the Red Cross in 2002. Sarah Harper in Oxford is quoted from her paper "Addressing the Implications of Global Ageing," *Journal of Population Research* 23 (2006): 205. Gavin Jones presented "Challenges of Ageing in East and Southeast Asia" at a conference called "Impact of Ageing: A Common Challenge for Europe and Asia" at the University of Vienna in June 2006 (www.univie.ac.at/impactofageing/pdf/jones.pdf).

Xiao Caiwei, international director of the China National Committee on Ageing, sent me a copy of his paper "The Ageing of Population and Its Implications in China" (www.unescap.org/esid/psis/meetings/Ageing_Change_Family/China.pdf), delivered in July 2007 at a seminar in Bangkok, "Social, Health and Economic Consequences of Population Ageing in the Context of Changing Families," organized by the UN Population Fund. Nicholas Eberstadt's "slow-motion tragedy" in China is discussed in his paper *China's Future and Its One-Child Policy* (Washington, DC: American Enterprise Institute, September 2007, www.aei.org/docLib/20070919_070918_Eberstadt_g.pdf).

Euthanasia in China was discussed in "Merciful or Ruthless? Plea for Euthanasia Chills China," *China Daily*, February 8, 2007 (www.china.org.cn/english/health/199437.htm). Margaret Sleeboom-Faulkner wrote on "Chinese Concepts of Euthanasia and Health Care" in *Bioethics* 20 (2006): 203. And Toshiko Kaneda looked at "China's Concern over Population Aging and Health" in a paper for the Population Reference Bureau in 2006 (www.prb.org/Articles/2006/ChinasConcernOverPopulationAgingandHealth.aspx).

Jesse Ausubel discussed the social effects of losing rebellious younger siblings, and much else, in the mind-bending "Human Population Dynamics Revisited," with Cesare Marchetti and Perrin Meyer, *Technological Forecasting and Social Change* 52 (1996): 1–30 (http://phe.rockefeller.edu/poppies), and in "Reasons to Worry about the Human Environment," *COSMOS* (Journal of the Cosmos Club of Washington, DC) 8 (1998): 1–12 (http://phe.rockefeller.edu/reasons-to-worry).

CHAPTER 27: SILVER LINING

A lot of good material on aging appears in "Global Aging: The Challenge of Success" by Kevin Kinsella and David R. Phillips, *Population Bulletin* 60, no. 1 (March 2005), www.prb.org/pdf05/60.1GlobalAging.pdf. John Bongaarts asks *How Long Will We Live?* in a report for the Population Council in December 2006 (www.popcouncil.org/pdfs/wp/215.pdf).

Guy Brown was writing about Helen Mirren and others in "The Long Game," *Guardian*, September 10, 2008. Katherine Whitehorn is quoted from her article "The Truth about Life after 80" in the same paper on August 29, 2008. Nicholas Eberstadt's quote on an absence of innovation from the old is from "Global Demographic Outlook to 2025," a lecture given in Zurich in November 2006 for the Progress Foundation's Economic Conference on Demography, Growth, and Wellbeing (www.progressfoundation.ch/PDF/referate/93_Lecture%20Nicholas%20 Eberstadt%20inkl.PPT_30.11.2006_E.pdf). "China's Contentious Pensioners" are analyzed by William Hurst and Kevin O'Brien in *China Quarterly* 170 (June 2002): 345 (http://polisci.berkeley.edu/faulty/bio/permanent/OBrien,K/ CQ2002.pdf).

I visited South Africa's AIDS grannies for HelpAge International (HAI) in 2002. I published interviews in the *Boston Globe* ("A Lost Generation," July 1, 2003) and in HAI's magazine *Ageing and Development* ("Stretching the Safety Net," January 2003).

Naohiro Ogawa was still in business examining the economic consequences of the longevity revolution in Japan in "Ageing in Japan: The Health and Wealth of Older Persons," a paper presented in Mexico City in September 2005 at the United Nations Expert Group Meeting on Social and Economic Implications of Changing Population Age Structures (www.un.org/esa/population/meetings/Proceedings_ EGM_Mex_2005/ogawa.pdf).

The benefits of an aging workforce are analyzed in a paper by Bo Malmberg et al., *Productivity Consequences of Workforce Ageing: Stagnation or Horndal Effect?* (Institute for Futures Studies, 2005, www.vxu.se/ehv/filer/forskning/cafo/semina-rie/Lindh.pdf). John MacInnes also takes an optimistic outlook in his 2005 paper with Julio Pérez, "The Reproductive Revolution" (see notes for chapter 14). David Bloom, Wolfgang Lutz, and Alexia Prskawetz and colleagues wrote about "Population Aging, Human Capital Accumulation and Productivity Growth" in a supplement to *Population and Development Review* 34 (2008). Lutz et al. also discussed "Future Ageing in Southeast Asia" in a paper for the International Institute for Applied Systems Analysis (Interim Report IR-07-026, www.iiasa.ac.at/Admin/ PUB/Documents/IR-07-026.pdf).

See Theodore Roszak's excellent books *The Longevity Revolution* (Albany, CA: Berkeley Hills Books, 2001) and *The Making of a Counter Culture* (Berkeley: University of California Press, 1969). He also wrote "Why Grey Matters" in *New Scientist*, August 31, 2002, and a letter published in the same magazine, August 10, 2002. Also see his "The Ecology of Wisdom" in *Whole Earth*, Fall 2002, and "Population Growth: No Longer a Problem" in the *Ecologist*, June 2002. Maggie Kuhn is quoted by Roszak in *The Longevity Revolution*.

CHAPTER 28: PEAK POPULATION AND BEYOND

Vaclav Smil vividly described the demographic turning points of the late twentieth century in "How Many Billions to Go?" in *Nature* 401 (1999): 429. Joel Cohen wrote on how future generations will look back at this time in "Human Population Grows Up," *Scientific American*, September 2005. See also his "World Population in 2050: Assessing the Projections" in *Seismic Shifts: The Economic Impact of Demo-*

graphic Change, edited by Jane Sneddon Little and Robert K. Triest (Federal Reserve Bank of Boston Conference Series No. 46 (June 2002): 83–113, www.bos.frb.org/economic/conf/conf46/conf46d1.pdf). Wolfgang Lutz looked at future population projections and other things in "From Population Explosion to Expanding Human Capital: Changing Challenges for the Future of Humankind," presented at the Club of Rome's International Conference on Strategies for a Sustainable Planet, June 16, 2008 (www.clubofrome.at/events/2008/rome/presentations/lutz.pdf).

For the idea of the baby boomers as the axis around which this story hangs, I am indebted to Richard Heinberg and his book *Peak Everything* (Forest Row, UK: Clairview Books, 2007). And for the news that the tsunami on December 26, 2004, left the planet with fewer people than it started the day, I thank Robert Engelman's *More: Population, Nature and What Women Want* (Washington, DC: Island Press, 2008). That day is likely to be the first of many thousands before the century is out.